Autonomous Flying Ad-Hoc Networks

Scrivener Publishing
100 Cummings Center, Suite 541J
Beverly, MA 01915-6106

Publishers at Scrivener
Martin Scrivener (martin@scrivenerpublishing.com)
Phillip Carmical (pcarmical@scrivenerpublishing.com)

Autonomous Flying Ad-Hoc Networks

Edited by

Taskeen Zaidi

Adarsh Kumar

Saurav Mallik

and

Keshav Kaushik

Scrivener Publishing

WILEY

This edition first published 2025 by John Wiley & Sons, Inc., 111 River Street, Hoboken, NJ 07030, USA and Scrivener Publishing LLC, 100 Cummings Center, Suite 541J, Beverly, MA 01915, USA

For more information about Scrivener publications please visit www.scrivenerpublishing.com.

Wiley Global Headquarters
111 River Street, Hoboken, NJ 07030, USA

For details of our global editorial offices, customer services, and more information about Wiley products visit us at www.wiley.com.

The manufacturer's authorized representative according to the EU General Product Safety Regulation is Wiley-VCH GmbH, Boschstr. 12, 69469 Weinheim, Germany, e-mail: Product_Safety@wiley.com.

Library of Congress Cataloging-in-Publication Data

ISBN 9781394231058

Cover image: Generated with AI using Adobe Firefly
Cover design by Russell Richardson

Set in size of 11pt and Minion Pro by Manila Typesetting Company, Makati, Philippines

Printed in the USA

10 9 8 7 6 5 4 3 2 1

Contents

Preface

The flying *ad hoc* network (FANET) is becoming popular as it involves unmanned aerial vehicles (UAVs), which can be useful in urban surveillance, disaster management, and rescue operations. The UAVs can be useful in various areas such as agriculture, surveillance, and telecommunications. The UAVs can be categorized by flight types, i.e. autonomous or remote, size, types of wings, and communication capabilities. The wings of UAVs are fixed and rotatory. The FANET is a subset of mobile *ad hoc* networks (MANETs), which are composed of mobile devices like laptops, cellular phones, and sensor devices. This also includes vehicular *ad hoc* networks (VANETs), which are composed of cars, buses, and ambulances based on vehicle-to-vehicle and vehicle-to-roadside infrastructure. The FANET shares some common characteristics of MANETs and VANETs, but nodes are more mobile in FANETs. The mobility degree of nodes in FANETs is higher than the mobility degree of nodes in MANETs and VANETs nodes. The MANETs or VANETs are walking people or vehicles on roads, but FANETs nodes fly in the sky. The topology changes frequently in FANETs, and FANETs create peer-to-peer networks for the coordination and collaboration of UAVs. In FANETs, a swarm of UAVs connect together locally with a base station in a nearby environment to get information. The UAVs can fly on a pre-programmed flight pilot or may be operated on complex automation systems, which provide versatility and flexibility during implementation and also communicate with environment and relays. The distance coverage for FANETs is larger than MANETs and VANETs. Mobile sensors were acquired by UAVs and all sensors possess different capabilities.

In this book, FANETs will be discussed in detail. The comparison between FANETs, MANETs and VANETs will be discussed first with the characteristics, features, and design models, and then the application areas of FANETs will be discussed covering their design issues and challenges. The existing FANET protocols will be discussed along with open research issues in the FANETs.

Chapters in this book are organized as follows:

Chapter 1 titled **"Research Perspectives of Various Routing Protocols for Flying *Ad Hoc* Networks (FANETs)"** discussed that FANETs are a specialized type of MANETs where the nodes are UAVs. Unlike single-UAV systems, only multi-UAV systems can form a FANET. These networks are gaining attention due to the rise in MANETs and the growing use of UAVs, especially in the military and public sectors. The FANETs face challenges in routing due to the high-speed mobility of UAVs. Effective routing algorithms are essential to ensure reliable communication among UAVs, making this a key area of research. Various routing strategies are being explored to improve data transfer performance in FANETs, with a focus on both their advantages and limitations.

Chapter 2 titled **"Exploring Quantum Cryptography, Blockchain, and Flying *Ad Hoc* Networks: A Comprehensive Survey with Mathematical Analysis"** introduced FANETs that are used in applications like disaster relief and environmental monitoring but face security challenges due to their mobility, dynamic topology, and resource constraints. This paper surveys security issues across various network layers, including physical layer threats like malicious drones and eavesdropping, data link layer vulnerabilities with secure transmission challenges, and network layer risks such as blackhole attacks. It also examines application layer concerns about data integrity and mission-critical security, as well as emerging threats like remote hijacking and GPS spoofing. The survey highlights the need for standardized security solutions and calls for collaborative research to address these challenges in FANETs.

Chapter 3 titled **"A Survey on Security Issues, Challenges, and Future Perspectives of FANETs"** studied that the recent advancements in digital and electronic systems have significantly improved the efficiency UAVs, particularly through the miniaturization and cost reduction of devices. The UAV networks, known as FANETs, are becoming increasingly common in both military and civil applications. A group of UAVs working together is more cost-effective and efficient than a single UAV system, prompting the development of new networking techniques for UAVs and ground control stations. However, to ensure FANETs function as reliable, robust, and context-specific networks, several challenges must be addressed. This chapter provides a comprehensive review of current communication infrastructures, security issues, and future perspectives. It discusses the security concerns, networking challenges, routing protocols, security requirements, mobility, and trajectory optimization models used to address UAV communication and collaboration issues, offering researchers an overview of key areas needing further exploration.

Chapter 4 titled **"Quantum Cryptography for Secure FANET"** discussed the essential security of FANETs due to their dynamic topology, high mobility, and decentralized nature. The UAV networks face various routing attacks, including sinkholes, blackholes, and data tampering. Current encryption methods, based on mathematical complexity, are vulnerable to quantum computers. Quantum cryptography offers a solution by leveraging the laws of physics to secure communication, providing data protection, and eavesdropping detection. Implementing quantum cryptography requires specialized hardware and techniques like post-quantum cryptography, the no-cloning theorem, and quantum key distribution to establish a shared secret key, making it resistant to man-in-the-middle attacks.

Chapter 5 titled **"A Review of Various Routing Protocols for FANET"** explored that FANETs are an emerging research area with increasing drone usage driven by advancements in sensors, processors, and applications. The FANETs are utilized in various fields, including disaster responses (like floods) and military operations. A major challenge for FANETs is dynamic routing, as the high speed of drones requires adaptive routing processes. This article explores the proposed solutions to address these routing challenges.

Chapter 6 titled **"The Integration of the Internet of Things in FANET"** discussed that the integration of the Internet of Things (IoT) with FANETs has the potential to transform various industries by enhancing data collection, sharing, autonomous task execution, traffic control, and providing valuable insights. The FANETs, consisting of UAVs or drones, can benefit from IoT in applications such as surveillance, disaster management, precision farming, and environmental monitoring. However, this integration poses challenges, including managing power consumption, ensuring secure communication, addressing privacy issues, and maintaining stable connectivity in dynamic environments. These factors must be carefully addressed for a successful IoT-FANET integration.

Chapter 7 titled **"Enhancing Precision Agriculture Through Bio-Inspired Routing Protocols for Flying *Ad Hoc* Networks"** discussed the integrating modern technology into agriculture, which is essential in meeting the needs of a growing global population, with precision agriculture being a key solution to optimize crop cultivation and reduce environmental impacts. However, challenges in connectivity and data acquisition in remote agricultural areas hinder its implementation. Integrating FANETs, which involve UAVs creating communication networks, offers a promising solution for real-time data collection and monitoring in agricultural fields. This integration, however, introduces complex routing issues due to the dynamic movement of UAVs, changing network structures, and energy constraints. The paper analyzes various routing protocols for efficient UAV communication, focusing on the application of bio-inspired protocols for FANETs.

Chapter 8 titled **"Disaster Recovery Management in FANETs"** observed that the FANETs are increasingly used in applications like aerial surveillance, disaster relief, and environmental monitoring but face vulnerabilities due to dynamic and unpredictable environments, such as communication failures and node crashes. Effective disaster recovery management is crucial to ensure FANETs remain operational during disruptions. This involves fault detection (e.g., node health monitoring, link quality assessment), fault tolerance (e.g., redundancy, cooperative communications, load balancing), and network reconfiguration (e.g., node relocation, route reestablishment). These strategies help maintain network performance and resilience. The use of cloud and edge computing can enhance recovery capabilities, and future research is focused on energy-efficient mechanisms, security, and integration with ground-based networks. Addressing these challenges will improve FANET reliability for critical applications.

Chapter 9 titled **"AI-Based Cybersecurity Opportunities and Issues on IIoT"** discussed that the rise of the Industrial Internet of Things (IIoT) has enhanced efficiency and automation but also introduced new cybersecurity challenges, putting critical infrastructure at risk. Artificial intelligence (AI) offers a potential solution by improving IIoT security through better threat detection, anomaly identification, and cyberattack prevention. This chapter explores the benefits and drawbacks of using AI in IIoT cybersecurity, including the risks of false positives/negatives, the need for constant model updates, and the vulnerability of AI to manipulation. The goal is to provide a comprehensive understanding of AI's potential in enhancing IIoT security while addressing its challenges and offering suggestions for its effective implementation in protecting vital infrastructure.

Chapter 10 titled **"Exploring the Synergy of Fog Computing and FANETs for Next-Generation Networking"** covered the integration of FANETs with fog computing and presents a promising approach for next-generation networking. Fog computing enhances FANETs by enabling real-time processing, reducing latency, and improving efficiency by bringing computing and storage closer to the network's edge. The FANET, consisting of autonomous UAVs, operates in a decentralized manner, and combining it with fog computing creates opportunities for applications requiring mobility, scalability, and distributed processing. This chapter explores the technical aspects of integrating these technologies, including communication protocols, resource management, and data processing strategies. It also discusses challenges like network connectivity, security, and UAV coordination, and highlights potential applications such as disaster management, surveillance, precision agriculture, and smart cities.

Chapter 11 titled **"Quantum Cryptography and FANET Security"** discussed various cryptographic techniques that ensure secure communication, but due to the open nature of wireless transmission, communication is often not considered secure. Quantum cryptography, which uses the principles of quantum physics, addresses these security issues by enabling secure information processing and transmission, especially in untrustworthy communication channels. The FANETs improve communication between ground stations and UAVs, supporting real-time data transfer while incorporating the security features of *ad hoc* networks.

The book offers an in-depth exploration of FANETs, focusing on their growing popularity and applications in areas such as urban surveillance, disaster management, rescue operations, agriculture, and telecommunications. The FANETs consist of UAVs that can be classified based on flight type (autonomous or remote), size, wing type (fixed or rotatory), and communication capabilities. The FANETs are a subset of MANETs and share similarities with VANETs, but with UAVs, nodes in FANETs exhibit higher mobility compared to MANETs or VANETs. The FANETs are characterized by rapidly changing topologies and local peer-to-peer communication between UAVs and base stations. The book compares FANETs with MANETs and VANETs, examining their characteristics, design models, and application areas. It also addresses FANET design challenges, existing protocols, and open research issues. This resource is essential for researchers, students, and professionals seeking to deepen their understanding of FANETs and their applications in modern network systems.

1

Research Perspectives of Various Routing Protocols for Flying *Ad Hoc* Networks (FANETs)

Kanthavel R.[1], Adline Freeda R.[2], Anju A.[2], Dhaya R.[1]* and Frank Vijay[3]

[1]*School of Electrical and Communication Engineering, The Papua New Guinea University of Technology, Lae, Papua New Guinea*
[2]*Department of Information Technology, KCG College of Technology, Chennai, India*
[3]*Department of CSE, SRM Eswari Engineering College, Chennai, India*

Abstract

Flying *ad hoc* networks (FANETs) are a novel type of mobile *ad hoc* networks (MANETs) in which the nodes are unmanned aerial vehicles (UAVs). This definition states that only multi-UAV systems may establish a FANET, while single-UAV systems are ineligible. However, not all multi-UAV systems come together to form a FANET. The industry has begun to pay more attention to FANETs as a result of the extensive use of MANETs. *Ad hoc* networks include routing technologies as a crucial component. The high-speed mobility of nodes makes FANET routing technology more difficult to implement. When conventional *ad hoc* networks or MANETs are unable to communicate, FANETs can gather data quickly and accurately. The FANET has recently gained popularity as a study topic due to the growth of UAV communication and is now widely employed in both military and public sectors due to its distinct advantages. The FANET is a unique class of networks made up of UAV nodes that can be utilized to carry out data transport under specific uncommon conditions. The crucial element and one that needs to be carefully constructed is a routing algorithm, which is necessary to provide robust and reliable communication among UAVs. This topic has drawn a lot of attention from researchers due to its significance and utility, and other routing strategies have also been proposed to enhance the caliber of data transfer in FANETs. The

**Corresponding author*: dhayavel2005@gmail.com

Taskeen Zaidi, Adarsh Kumar, Saurav Mallik and Keshav Kaushik (eds.) Autonomous Flying Ad-Hoc Networks, (1–14) © 2025 Scrivener Publishing LLC

performance of several routing protocols in real time is surveyed in this chapter. This study is based on the benefits and drawbacks of several FANET protocols.

Keywords: Flying *ad hoc* networks, mobile *ad hoc* networks, unmanned aerial vehicles, routing protocols

1.1 Introduction

Flying *Ad Hoc* Networks (FANETs) are composed of unmanned aircraft capable of connecting with end nodes. At any given time, at least one aircraft must remain connected to a ground control station or designated location. Aerial vehicles, like automatic pilots, function without human assistance [1]. This is due to a new focus on FANETs in various research domains, like universities and companies, due to the availability of more affordable and compact wireless communication devices. A subdomain of mobile *ad hoc* network (MANETs) mentioned as FANETS employs a collection of homogenous flying objects called micro air vehicles (MAVs) that connect locally. Emergency situations, catastrophes of nature, military engagement zones, and unmanned airliners can all be supported by FANETs. When conventional *ad hoc* networks or MANETs are unable to communicate, FANETs can gather data quickly and accurately. Today, FANETs are utilized in a variety of military and civilian applications, including managing wildfires [2] and monitoring disasters [3]. It is crucial to adopt trustworthy protocols for these types of systems and evaluate their model capability since every category of associations possesses specifications, and the use of the protocol relies on these specifications. Mobility models and communicating traffic patterns are two aspects that impact protocol modeling, among others.

A FANET is a system of little unmanned aerial vehicles (UAVs) associated in an *ad hoc* method, working together to accomplish high-level objectives. The major advantages of FANETs, which could increase their connections and widen their communication range at infrastructure-less locations, are their accessibility, absence of centralized management, and individual manageability. They also have less infrastructure compared to other UAVs [1]. On one end of the spectrum, FANETs can be employed to deliver a quick arrangement. They are composed of adaptable auto-configurable devices that have a comparatively small cost of ownership network in critical circumstances when common communication capacity is not accessible; in some cases, associating more UAVs in an infrastructure-less network has been a typical test. To achieve an efficient and resilient connection at this degree of collaboration, the right network configuration protocols must be configured on extremely rapidly flying components.

The remainder of this chapter is structured as following: Unmanned aerial vehicle is provided in Section 1.2. The properties of FANET are presented in Section 1.3. The primary FANET routing protocols are shown in Section 1.4. In Section 1.5, the FANETs' communication pedagogy is illustrated. In Section 1.6, the difficulties with FANET configuration are discussed, and the concluding remarks bring this study to a close [2].

1.2 Unmanned Aerial Vehicles

Drones, mentioned as UAVs, are extensively utilized in manufacturing, including telecommunications, agriculture, monitoring wildfires, and border patrol. Modest classifications can be informed to classify UAVs, for example, according to their size (big or tiny), such as the sort of the wings, the mode of fluttering (autonomous or remotely controlled), or their ability for communication. Unmanned aerial vehicles with rotors can have either stationary flaps or rotary flaps. The rotary wing UAVs can perform a horizontal takeoff and landing. They are stable (they can control yaw, diameter, move, and acceleration) and will hover above stationary objects [3]. Fixed wing UAVs, in comparison, have more aerodynamics designs, quicker soaring velocity, and longer flight times. In terms of interaction, we differentiate both single UAVs and multi-UAV platforms. Star structures are frequently used by single UAVs to connect to a satellite. Most military and civil uses currently are executed using multi-UAV structures made up of small UAVs [4]. Shorter completion times, lower expenses, more flexibility, and a greater reliability are just a few advantages to this tactic, also known as a FANET.

A thorough classification of UAVs is given based on a number of variables, including the dimensions and mass of the UAV, its altitude, its flight distance, intended use, flying procedure, ownership, airspace class, degree of piloting independence, and engine type [5]. These variables are all depicted in Figure 1.1. When choosing which UAV is the best for a mission, the size is usually a crucial consideration. The UAVs can be categorized into three categories, as seen in Figure 1.1.

The UAVs based on their flight mechanics is shown in Figure 1.2. Fixed-wing UAVs that may be flown remotely by the pilot or autonomously by onboard computers can use wings to fly similarly to airplanes. At least four rotors are required for multi-rotor UAVs to maintain flight, while hybrid variants combine the advantages of the two. Box F: Classification by airspace, which is based on airspace categorization, is another way to categorize UAVs [6].

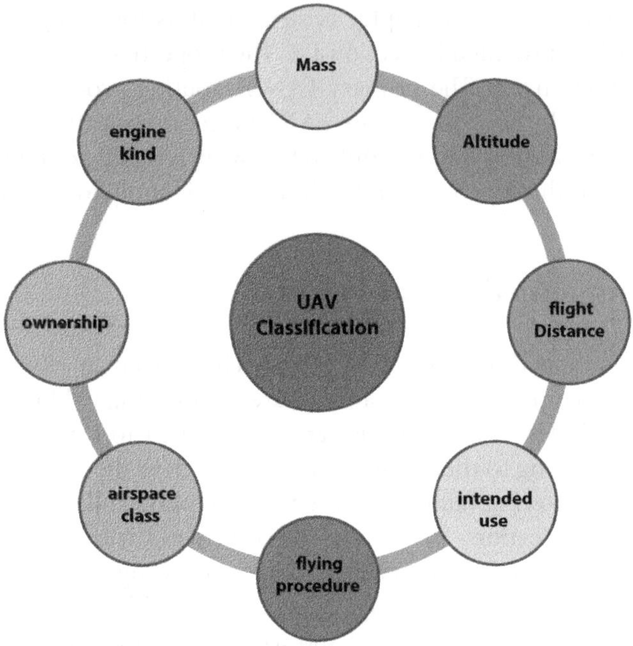

Figure 1.1 UAV classification – structures.

Flying Mechanism of UAV		
Fixed-wing drones	Multi-rotor UAVs	Hybrid fixed/rotary wing UAVs

Figure 1.2 UAV's categorizing.

According to the level of autonomy in their flight procedures, UAVs are divided into two primary groups by the International Civil Aviation Organization (ICAO), namely completely self-driving and autonomously piloted aircraft. According to their ownership, as indicated in Box H, UAVs can be categorized as either private or public, i.e., whether they are used and possessed by community organizations, such as regional regulation forces and government organizations, or by the private sector [7].

In accordance with Box I, UAVs can also be divided into groups depending on the kind of propellers they possess. Particularly for tiny UAVs and heavy UAVs, respectively, piston and battery-powered engines are the most often used propulsion types.

1.3 FANET Characteristics

There are networks made entirely of aerial devices called FANETs, which may communicate between UAVs.

The UAV industry is quickly growing. It is utilized for intelligence gathering, route research, security inspection, target monitoring, military strike, and other activities due to its compact size, adaptable movement, and lengthy flight time. Multiple UAVs must cooperate in sophisticated applications, though. Multiple UAVs can independently build a cooperative integrated communication system known as a "flying *ad hoc* network" to carry out more difficult network activities [8].

Every UAV in a FANET functions as a router, and the wireless link that connects them creates a network for the purpose of relay communication. The position of UAV nodes influences throughput, an important characteristic of system outcome.

By integrating UAVs into a FANET, it becomes possible to dynamically adjust the link between two nodes. Optimizing connectivity enhances overall network throughput.

If land-based networks are congested or unavailable, FANETs can assist other types of networks, such as cellular or satellite networks, by sending the flow received from such devices to a distant server [9]. Figure 1.3 shows the FANET characteristics.

Node intensity: The median number of UAVs per region is known as the node's intensity. The FANETs have a fewer UAV node count than other *ad hoc* networks like MANET and vehicular *ad hoc* networks (VANET).

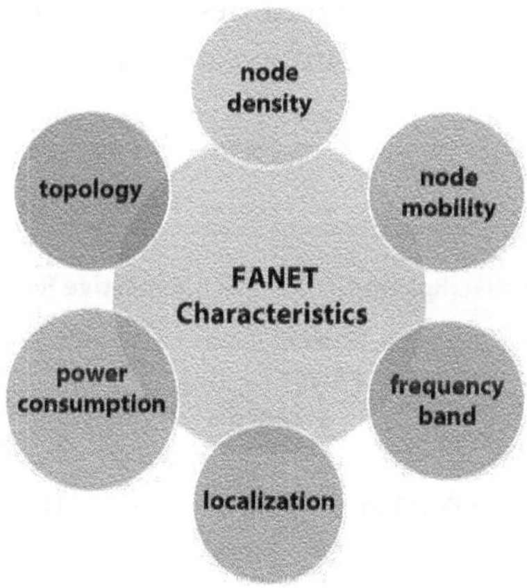

Figure 1.3 FANET characteristics.

Network movement: Network movement, which is significantly greater in FANETs than in VANETs and MANETs, is one of their important properties. Unmanned aerial vehicles travel so swiftly that FANET structures routinely alter as a result. Potential FANET configurations, according to situations of continuous topographical instability, contain one type of topology (star) directly connected through the land managing center and a network arrangement for the continuous direction-finding requirement.
Localization: The FANETs use low-level global positioning (GPS) to find UAVs because they are swift and mobile. They alter network topology. In FANETs, localized methods include network positions, elevation assistance GPS (AGPS), and differentiated GPS (DGPS).
Power usage: *Ad hoc* networks struggle greatly with energy limitations. The amount of power needed in FANETs is influenced as a result of the dimension of the UAVs, the travel differences, the technology for commuting utilized in the FANET, the connection, and other barriers.
Frequency band: 0.9 GHz and 2.4 GHz are two unlicensed bands that are frequently used in UAV communication systems. Congestion can arise from using these bands, though. For UAV-to-ground communications, IEEE 802.11a combined with a 5 GHz frequency yields the greatest results [10].

1.4 Routing Protocols for FANETs

When using UAVs to deliver data, a procedure called routing is used to determine the most effective route. An efficient path has a high delivery ratio, fewer relay nodes, less control messages, and so on. It also reduces delivery times. Because of their unique qualities and the environment in which they operate, FANETs should design their own routing methods. The FANET routing protocols can be categorized by four major types, as shown in Figure 1.4.

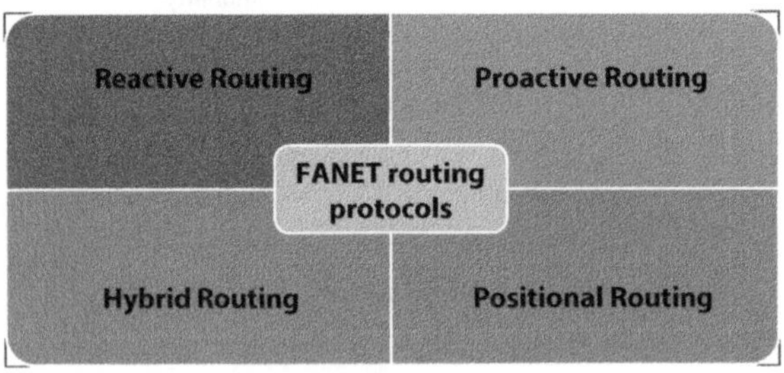

Figure 1.4 The major categories of FANET routing protocols.

Reactive routing protocol (RRP): It is referred to as an "on-demand routing protocol." The route between the two nodes do not need to be stored if there is no communication between them. There are two distinct message types [11], which are route replies and route request messages. Source nodes generate route request messages, and destination nodes generate route replies. In extremely dynamic networks, this form of routing protocol has low stability and poor security, such as dynamic source routing (DSR) and *ad hoc* on-demand distance vector.

Proactive routing: Tables are used by proactive routing protocols (PRP) to accumulate the steering information for a node or nodes in a certain network region belonging to each other.These tables enable efficient data forwarding, even during topology changes, due to the use of various table-driven mechanisms employed in FANETs. Proactive routing's key benefit is that it has the most recent data on the routes, making it simple to choose a pathway from the source nodes to the beneficiary nodes, eliminating the need to delay.

Hybrid routing: The hybrid routing protocol (HRP) was developed. With RRPs, the initial route discovery process could have a high latency; with PRPs, the delay of control information could be reduced.

Positional routing: Consuming GPS, each node in the position based routing protocol determines its own location and broadcasts it to its neighbors along with its current location and speed. Next, the source determines the location and speed of the destination and creates a region by determining the intermediate node with the minimum distance to the endpoint nodes. This allows the source to select which routing strategy to employ because different approaches have different requirements.

1.5 Communication Pedagogy for FANETs

It is critical to establish dependable communicating processes for UAV setups because communication is one of the FANET's most important design considerations. The UAVs and ground control stations (GCS) make up the FANET architecture, which are composed of a swarm of little flying objects. The UAVs are equipped with cameras, GPS, and sensing. The UAVs provide long-distance, line of sight (LOS), and environment-resistant communication. The UAVs are set up in low, middle, and elevated levels, according to altitudes. Figure 1.2 above depicts the high altitudes platform-based FANET design.

To establish a U2U link with other UAVs and collect network data, the UAV also develops a U2I link with the infrastructure's ground control station. The UAVs are referred to be member UAVs and backbone UAVs depending on these responsibilities. Data is gathered from member UAVs and transmitted to GCS by the backbone UAV.

To connect several UAVs, there are primarily two types of FANET architectures: centralized and decentralized is shown in Figure 1.5.

- **Centralized architecture:** The GCS serves as the hub of this FANET system, to which all other UAVs are directly connected. Thus, the GCS is used to transmit data between any two UAVs.
- **Decentralized architecture:** The UAVs can interact both directly and informally in this kind of architecture. Three subtypes of the decentralized architecture are further separated into multilevel-group UAV *ad hoc* networks, and multilevel-layer UAV *ad hoc* networks constitute three further variants of the decentralized structure [12].

The FANET uses the following methods of communication.

- **Inter-plane communication:** The communication between UAVs of two different planes.
- **Intra-plane communication:** The communication between UAVs that are on the same aircraft.
- **Ground station communication:** A plane's UAVs' ability to communicate with a base station or ground station.
- **Ground sensor communication:** The interaction of any aircraft's UAVs with ground-based sensors.

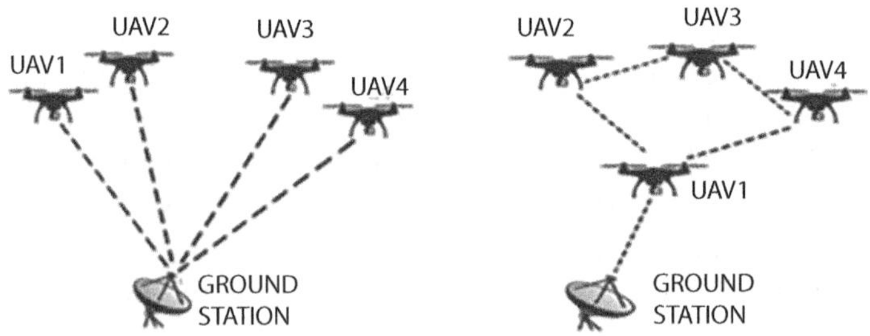

Figure 1.5 Centralized and decentralized architecture.

1.6 Challenges and Applications of FANET Configuration

We could build a UAV system that could flutter independently or be tenuously controlled because of rapid advancements in electronic devices and communication technologies [13]. The use of UAVs ensures modern methodologies for both armed and civilian solicitations due to their ease of operation (adaptability, flexibility, and equipment accessibility). Figure 1.6 elaborates the challenges in FANETs.

1.6.1 Issues and Challenges in FANETs

- **National laws:** In various FANET application areas, the UAV is used. Current air regulations in several nations prohibit restricted UAV operations in civil airspace.
- **Routing:** Major issues with data routing between UAC development of novel routing algorithms and networking modals is necessary to construct the flexible routing conventions, which ought to be ready to change the routing databases in accordance with topology updates.
- **Path planning:** In FANET, path planning plays a crucial role. Therefore, learning new algorithms, techniques, and dynamic path planning is crucial to achieving the objectives. Many times, each UAV must alter its previous course, and fresh ones must be animatedly planned.

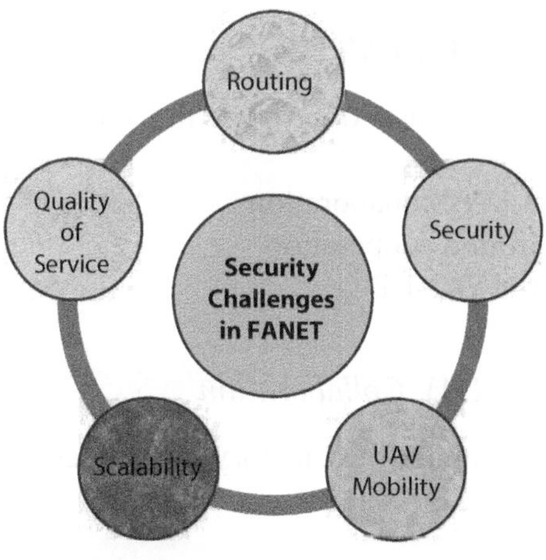

Figure 1.6 Challenges in FANETs.

- **Collaborating of UAVs and manned aircraft:** To maintain the FANET aircraft network environment and to complete the flying requirement, UAV collaboration is required. The coordination of UAV will allow the smallest possible loss of attacking planes.

Routing: Since there are high node movements and frequent topology changes in FANETs, routing is different from that of other *ad hoc* networks. Routing algorithms should have great mobility and can update rapidly.
Security: The secure routing point for managing FANETs is to guarantee the availability, confidentiality, and integrity of sensitive data. However, a problem in FANETs is that the nodes' physical security can be compromised. Another critical point is trust management. Nodes in FANETs join and exit somewhat often. *Ad hoc* network accessibility routing methods cannot withstand hostile assaults and frequent network topology changes in FANET [14].
Unmanned aerial vehicle mobility: The deployment of UAVs is the primary concern in FANET since they have a variety of characteristics and capacities that may be used for various purposes. Optimizing UAV placement is a current challenge.
Scalability: Since a single UAV can only accomplish a particular task, several UAVs are needed to complete the tasks related to perfume. To accommodate several UAVs, the FANET algorithm design should be flexible.
Quality of service: It is necessary to raise the parameters of service quality. In FANET, many types of data, including images, videos, audio files, and texts, are changed.

1.6.2 Applications for FANETs

The following are the key uses of FANETs that could be identified:

- Multilevel-UAV collaboration
- UAV-to-ground cooperation
- UAVs in the VANET

1.6.2.1 Multilevel-UAV Collaboration

Applications that fall under the category of multilevel-UAV collaboration include the ones below:

- Targeting recognition: UAVs may employ target recognition technological advances, such as infrared and visual photographic equipment, to find individuals and things.
- Surveillance and tracking during situations of emergency: UAVs can help determine the path of flooding and subsequently predict which buildings will be damaged. According to this, they may be utilized to prioritize rescue operations after earthquakes by identifying destroyed structures with significant populations, such as medical centers and educational institutions, and giving these places a greater importance.
- In critical circumstances, UAVs are used in the development sector to keep tabs on work progress and evaluate quality. Besides many other emergencies, UAVs can be used to briefly boost wireless coverage in cellular networks when the terrestrial ground stations are unavailable.
- Examples of remote-sensing FANET applications include multi-UAV collaboration, UAV-to-ground, and UAV-to-vehicular *ad hoc* networks [15].

1.6.2.2 UAV-to-Ground Cooperation

- Civilian and public solicitations: UAVs are frequently utilized in the public and civilian domains, particularly in the form of small quad copters, because they are more flexible and affordable than ground-based infrastructure.
- Rescue efforts: UAVs are essential to search and rescue (SAR) operations. When it comes to ensuring public safety, conducting SAR operations, the management of both natural and synthetic catastrophes, including quakes, hurricanes, wildfires, tsunamis, and terrorist attacks, as well as the assessment of the safety of critical infrastructures, such as facilities for water and electrical power, are all important tasks.
- The FANETs are thought to be a huge asset. In these circumstances, it is crucial to have communications coverage. The UAVs can deliver early catastrophe warnings and aid in accelerating rescue and recovery efforts in circumstances where public communications networks are down. The UAVs can transport medical supplies to remote locations. They can speed up SAR operations in a variety of circumstances, including avalanches, wildfires, and more.

1.6.2.3 UAVs in VANETs

The UAVs and automobiles work together in the ensuing applications:

- **Tracking of highway traffic:** The FANETs can replace time-consuming procedures and complex infrastructures for surveillance. When employed for tracking road traffic, UAVs can quickly identify crashes to authorities. Expending UAVs is significantly quicker than expending the accident's commandant's automobile. The UAVs could also be utilized to capture in-flight videos of various safety incidents and conditions on road systems to increase safety while driving.
- **Transmission of information packets:** Providing information to MANETs is a thought-provoking job in identifying a reliable relaying technique to confirm that information is distributed from one operator to another. The UAVs are commonly used in these circumstances to transmit information obtained by ground-based technologies to far-off control the center *via* airborne telecommunications' relaying. Using the load-carry-and-delivery framework, which depends on numerous UAVs to gather the data from the origin nodes and distribute it to the destination networks [16], UAVs transmit information in packs.
- **Poor path guiding:** It is a consequence of the increased mobility of cars in VANETs. Multi-hop relays are utilized in VANETs with UAV support to transfer information among automobiles and UAVs. The UAVs are used in VANETs to improve navigation and offer route directions.

1.7 Conclusion

Along with this effort, we examined the current state of the art for concepts of FANET mobility, placement, and transmission. The article begins with an overview of FANETs, moves on to a discussion of FANETs' properties, and concludes with a look at FANETs' protocols for routing. Furthermore, the hardest design problems for multi-UAV platforms are those involving interactions and coordination. The FANET, a distinct network family, is examined in this research as an *ad hoc* network connecting UAVs. Our conversation of potential UAV system applications that was produced feasibly by a survey of the key developing knowledge that is assimilated with FANETs is a significant contribution of this chapter. The future of this study is clarified by this conversation, which may also pave the way for fresh approaches to the issues raised here.

References

1. Rovira-Sugranes, A., *et al.*, A review of AI-enabled routing protocols for UAV networks: Trends, challenges, and future outlook, *arXiv preprint arXiv:2104.01283*, 2021.
2. Khan, M.A., *et al.*, Dynamic routing in flying *ad-hoc* networks using topology-based routing protocols. *Drones*, 2, 3, 27, 2018.
3. Yang, H. and Liu, Z., An optimization routing protocol for FANETs. *EURASIP J. Wirel. Commun. Netw.*, 2019, 1, 1–8, 2019.
4. Bekmezci, I., Sahingoz, O.K., Temel, Ş., Flying *ad-hoc* networks (FANETs): A survey. *Ad Hoc Netw.*, 11, 3, 1254–1270, 2013.
5. Barrado, C., Messeguer, R., López, J., Pastor, E., Santamaria, E., Royo, P., Wildfire monitoring using a mixed air-ground mobile network. *IEEE Pervasive Comput.*, 9, 4, 24–32, 2010.
6. Maza, I., Caballero, F., Capitán, J., Martínez-De-Dios, J.R., Ollero, A., Experimental results in multi-UAV coordination for disaster management and civil security applications. *J. Intell. Rob. Syst.*, 61, 1–4, 563–585, 2011.
7. Mozaffari, M., Saad, W., Bennis, M., Nam, Y.H., Debbah, M., A Tutorial on UAVs for Wireless Networks: Applications, Challenges, and Open Problems. *IEEE Commun. Surv. Tutor.*, 21, 2334–2360, 2019.
8. Chriki, A., Touati, H., Snoussi, H., Kamoun, F., FANET: Communication, mobility models and security issues. *Comput. Netw.*, 163, 106877, 2019.
9. Sahingoz, O.K., Networking Models in Flying *Ad-Hoc* Networks (FANETs): Concepts and Challenges. *J. Intell. Rob. Syst.*, 74, 513–527, 2014.
10. Sultan, L., Anjum, M., Rehman, M., Murawwat, S., Kosar, H., Communication Among Heterogeneous Unmanned Aerial Vehicles (UAVs): Classification, Trends, and Analysis. *IEEE Access*, 9, 118815–118836, 2021.
11. Lahmeri, M.A., Kishk, M.A., Alouini, M.S., Artificial Intelligence for UAV-Enabled Wireless Networks: A Survey. *IEEE Open J. Commun. Soc.*, 2, 1015–1040, 2021.
12. Zhang, D., Xu, F., Pun, C.M., Yang, Y., Lan, R., Wang, L., Li, Y., Gao, H., Virtual Reality Aided High-Quality 3D Reconstruction by Remote Drones. *ACM Trans. Internet Technol.*, 22, 1–20, 2021.
13. Euchi, J., Do drones have a realistic place in a pandemic fight for delivering medical supplies in healthcare systems problems? *Chin. J. Aeronaut.*, 34, 182–190, 2021.

14. Nyaaba, A.A. and Ayamga, M., Intricacies of medical drones in healthcare delivery: Implications for Africa. *Technol. Soc.*, 66, 101624, 1–8, 2021.
15. Pulsiri, N. and Vatananan-Thesenvitz, R., Drones in emergency medical services: A Systematic literature review with bibliometric analysis. *Int. J. Innov. Technol. Manag.*, 18, 2097001, 2021.
16. Chen, H., Lan, Y., Fritz, B.K., Hoffmann, W.C., Liu, S., Review of agricultural spraying technologies for plant protection using unmanned aerial vehicle (UAV). *Int. J. Agric. Biol. Eng.*, 14, 38–49, 2021.

2

Exploring Quantum Cryptography, Blockchain, and Flying *Ad Hoc* Networks: A Comprehensive Survey with Mathematical Analysis

Tarun Kumar Vashishth*, Vikas Sharma, Kewal Krishan Sharma, Bhupendra Kumar, Sachin Chaudhary and Rajneesh Panwar

School of Computer Science and Applications, IIMT University, Meerut, U.P., India

Abstract

Flying *ad hoc* networks (FANETs) have garnered significant attention for applications such as disaster relief, environmental monitoring, and remote communication. However, FANETs' mobility, dynamic topology, and resource constraints introduce security challenges. This paper surveys security issues in FANETs, covering vulnerabilities, threats, and countermeasures across various network layers. At the physical layer, threats like malicious drones, jamming, and eavesdropping are discussed, along with authentication mechanisms to prevent unauthorized access. The data link layer analysis includes communication link vulnerabilities and secure transmission challenges, reviewing encryption, key management, and intrusion detection methods. The network layer addresses routing security, exploring threats like black hole attacks and secure routing protocols. Application layer concerns focus on data integrity and mission-critical application security. Additionally, emerging challenges with unmanned aerial motors and autonomous systems, such as remote hijacking and GPS spoofing, are examined. The survey highlights open research areas and calls for collaborative efforts to develop standardized FANET security solutions.

***Keywords*:** Flying *ad hoc* networks, FANET, UAV, autonomous aerial systems, threats, vulnerabilities, mobility, dynamic topology

**Corresponding author*: tarunvashishth@gmail.com

Taskeen Zaidi, Adarsh Kumar, Saurav Mallik and Keshav Kaushik (eds.) Autonomous Flying Ad-Hoc Networks, (15–36) © 2025 Scrivener Publishing LLC

2.1 Introduction to Quantum Cryptography

Quantum cryptography (QC), a groundbreaking fusion of quantum mechanics and cryptography, represents a paradigm shift in secure conversation. It exploits the concepts of quantum physics to create encryption strategies, which are theoretically impervious to traditional cryptographic attacks. This phase offers a complete introduction to QC, setting up the foundational ideas and highlighting its transformative capability inside the realm of statistics safety.

2.1.1 Quantum Mechanics Primer

Quantum mechanics, the theoretical framework governing the behavior of particles at the atomic and subatomic stages, introduces several pivotal ideas that underpin quantum cryptography:

Quantum states: Unlike classical bits, which exist as either 0 or 1, quantum bits (qubits) can exist in a superposition of each state simultaneously. This duality paperwork is the idea for quantum facts processing.
Heisenberg's uncertainty principle: This essential precept asserts that certain pairs of residences, like the role and momentum of a particle, cannot be exactly measured simultaneously. Instead, there is inherent uncertainty, resulting in a probabilistic nature that QC exploits.
Entanglement: Quantum entanglement is a phenomenon in which the states of two or more particles end up interdependent, no matter the spatial separation. Changes to at least one particle immediately influence the other, even at sizeable distances, defying classical notions of locality.

2.1.2 Genesis of Quantum Cryptography

The origin of QC may be traced to the inventive paintings of Stephen Wiesner within the 1960s, who added concepts like "quantum money." However, it became the landmark 1984 paper *via* Charles Bennett and Gilles Brassard that laid the foundations for modern QC.

2.2 Quantum Key Distribution

Quantum key distribution (QKD) is an innovative and cryptographic technique that exploits the principles of quantum mechanics to establish a stable and unbreakable shared key between two parties (e.g. Alice and Bob, even with potential eavesdroppers). Quantum key distribution ensures the safety of the important alternate system by leveraging the precise properties of quantum debris with photons and the concepts of quantum mechanics, such as the Heisenberg uncertainty principle and the no-cloning theorem.

2.2.1 Components of Quantum Key Distribution

Photon source: The process begins with Alice generating a stream of individual photons, typically using a laser or other controlled light sources. These photons serve as carriers of quantum information.
Quantum state preparation: Alice encodes her secret information (the bits of the key) onto the quantum states of the photons. These states can be polarizations (e.g., horizontal and vertical), phases, or other quantum properties.
Quantum transmission channel: Alice sends the encoded photons to Bob through a quantum channel, a fiber-optic cable or unfastened space. This channel introduces noise and capacity disturbances to the quantum states.
Quantum measurement: Upon receiving the photons, Bob performs measurements on them using suitable measurement bases. The measurement bases are pre-arranged between Alice and Bob and are kept secret until the end of the protocol.
Key agreement: After performing measurements, Bob communicates with Alice over a classical communication channel to disclose the bases he used for the measurement (but not the results). Alice then reveals which bases she used for encoding. Figure 2.1 describes the basic elements of Quantum Key Distribution (QKD) systems, from photon emission to privacy amplification.

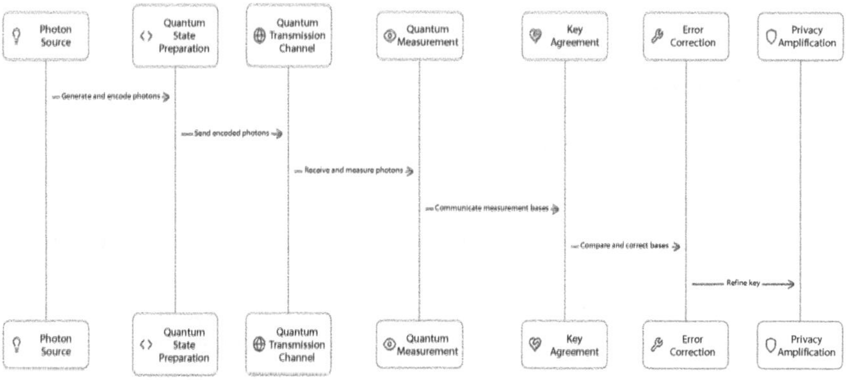

Figure 2.1 The QKD components with their links.

Error correction: Alice and Bob compare the bases they used. If they used the same bases, their measurements would result in a valid key bit. If not, those bits are discarded. This step corrects errors introduced by the channel.
Privacy amplification: To ensure security, Alice and Bob further refine their key by applying privacy amplification techniques. This process involves hashing and manipulating the key bits to reduce the possibility of an eavesdropper having any information about the final key.

2.2.2 Key Aspects and Security

Heisenberg's uncertainty principle: The QKD relies on the fact that measuring quantum states disturbs them. Any eavesdropping attempt by an adversary, Eve, will inevitably alter the quantum states, making her presence detectable.
No-cloning theorem: The no-cloning theorem of quantum mechanics prevents an eavesdropper from making perfect copies of the quantum states sent by Alice. Therefore, even if Eve intercepts the photons, she cannot replicate the exact key.

2.2.3 Challenges and Practical Considerations

Photon loss and noise: Quantum states are delicate and susceptible to losses and noise introduced by the channel and equipment, reducing the effective range of QKD.
Detection efficiency: Efficient detectors are crucial to minimizing errors and ensuring the successful implementation of QKD.
Post-processing complexity: Error correction and privacy amplification can be computationally in-depth, impacting the rate of key era.

2.2.4 Applications

The QKD has potential applications in various fields, including:

Secure communication: The QKD provides a means for ultra-secure communication between parties, ensuring that the exchanged key remains a secret.
Quantum cryptography networks: The QKD can be integrated into larger quantum networks for secure communication over longer distances.

2.3 Literature Review

Hafeez *et al.* (2023) [1] provided a complete evaluation of the mixing of privateness and security aspects within blockchain-assisted unmanned aerial motor conversation.

Hassija *et al.* (2021) [2] presented a detailed review of protection-vital drone applications and the associated demanding situations in drone communication, consisting of denial-of-service (DoS) assaults, man-in-the-middle assaults, de-authentication attacks, and more.

Wu *et al.* (2019) [3] conducted a comprehensive survey, categorizing blockchain technologies into four layers and exploring consensus strategies, network aspects, and blockchain applications, with a focus on areas like the Internet of Things (IoT).

Abdulhae *et al.* (2022) [4] offered a comprehensive survey of cluster-primarily based routing protocols (CBRPs) for flying *ad hoc* networks (FANETs), analyzing their strengths, weaknesses, packages, strategies, node matter, and ability future enhancements.

Ali *et al.* (2021) [5] gifted G-OLSR an overall performance-conscious routing mechanism designed to facilitate green communique and collaboration among unmanned aerial automobiles within a FANET context.

Rovira–Sugranes *et al.* (2022) [6] conducted a complete assessment of artificial intelligence (AI)-enabled routing protocols tailored for aerial networks, encompassing topology-predictive and self-adaptive getting-to-know-based routing algorithms, with a focal point on addressing the demanding situations posed *via* dynamic network topologies.

Khan *et al.* (2020) [7] introduced a modern routing protocol for FANET the use of a changed AntHocNet. Their observation employs ant colony optimization (ACO), showcasing enhanced dependability and performance in comparison to legacy course selection techniques. The introduction of an energy stabilizing parameter enhances electricity performance and network overall performance. Simulation results show the prevalence of the proposed protocol over universal ACO and different traditional FANET routing protocols.

Pasandideh *et al.* (2023) [8] spoke about the exploration of integrating rising technologies with FANETs and complications on the simulation tools used to assess the overall performance of FANETs.

Kumar *et al.* (2021) [9] analyzed modern research trends in quantum computing, quantum-secure computing, and put up-quantum cryptography applicable to quantum networks. The paper additionally highlights the benefits, constraints, potential advancements, and studies demanding situations associated with quantum technologies, drones, and their networks.

Liu *et al.* (2020) [10] gifted the pioneering achievement of cellular entanglement distribution *via* drones, allowing operations in numerous weather conditions together with sunlight hours and rainy nights. The Clauser–Horne–Shimony–Holt S-parameter is measured at 2.41 ± 0.14 and 2.49 ± 0.06 for these scenarios. This system's terrific mobility, flexibility, and reconfigurability provide wonderful blessings over satellite and fiber-primarily based quantum communication, highlighting its ability to establish a scalable multinode quantum community making use of symmetrical lens diameter and unmarried-mode-fiber coupling.

Fernandez–Carames *et al.* (2020) [11] provided an exploration of the cutting-edge country of put up-quantum cryptosystems and their ability utility to blockchains and distributed ledger technologies.

Wang *et al.* (2019) [12] conducted a complete analysis of cryptographic primitives in blockchains by means of reading the pinnacle-30 mainstream cryptocurrencies, analyzing their usages, functionalities, and evolutionary elements.

Longo *et al.* (2022) [13] advised a unique method for coping with cryptographic primitives in clever-contract-based total ledgers, addressing its applicability in each permissioned and absolutely permissionless settings.

Bhatt, Alekha Parimal, and Anand Sharma (2019) [14] conducted an evaluation of usual safety threats in IoT and explore current answers together with their barriers.

Islam *et al.* (2021) [15] presented an article outlining a blockchain-based, AI-empowered pandemic situation supervision scheme. The scheme employs a fleet of AI-prepared drones to autonomously screen and detect pandemic outbreaks, minimizing human intervention.

2.4 Blockchain Technology: Enhancing Security and Transparency

Blockchain generation, a floor-breaking innovation, holds the energy to reshape industries through heightened protection and transparency. Acting as a decentralized and tamper-evidence virtual ledger, blockchain transacts securely and transparently. This transformative generation offers multifaceted benefits, positioning it as a pivotal pressure throughout sectors like finance and delivery chain management. At its coronary heart, blockchain ensures data integrity through cryptographic hashing

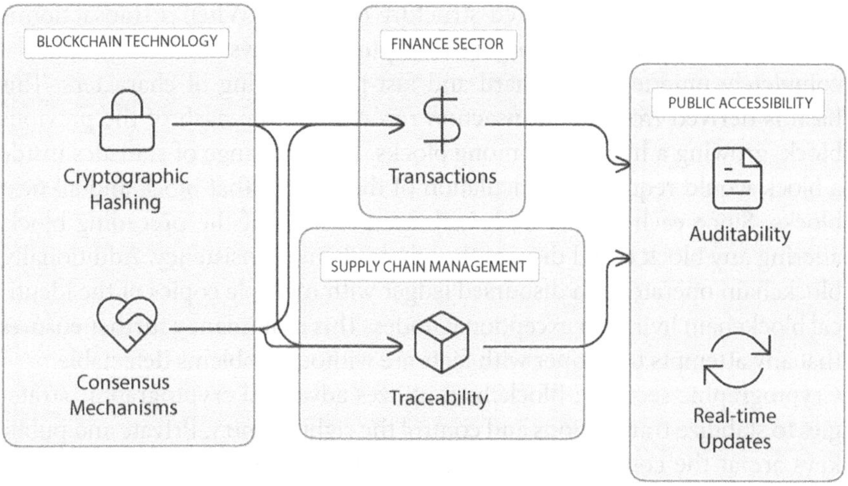

Figure 2.2 Cloud-based architecture of blockchain technology.

and consensus mechanisms. Its decentralized nature eradicates the need for intermediaries, ensuring belief without crucial management. Public accessibility to records on the blockchain fosters actual-time updates and auditability, reducing disputes and improving reliability. From cryptocurrencies to deliver chain traceability, blockchain's potential to save you from fraud, beautify performance, and raise transparency heralds a new generation of reliable and green systems in the current digital panorama. Figure 2.2 describes the functionality of blockchain technology within the cloud and its application in finance and supply chain management along with its traceability and real-time updating features.

2.4.1 Decentralization and Consensus Mechanisms

Blockchain operates on a decentralized network of computers (nodes), putting off the want for a government. Transactions are grouped into blocks and introduced into the chain *via* consensus mechanisms, including Proof of Work (PoW) or Proof of Stake (PoS). These mechanisms ensure that transactions are verified and brought to the blockchain while a consensus is reached among taking part nodes, ensuring consideration without intermediaries.

2.4.2 Enhancing Security

Immutability: Immutability is a foundational concept in blockchain technology that ensures data is immediately brought to the blockchain; it cannot be modified or deleted. This is done through an aggregation of cryptographic hashing and an interconnected structure of blocks. When a transaction is delivered to a block, it undergoes a cryptographic system that generates a completely unique hash, a hard and fast-period string of characters. This hash is derived from the transaction records and the hash of the previous block, growing a hyperlink among blocks. Any exchange of statistics inside a block would require a recalculation of the hash of that block and all next blocks. Since each block's hash includes the hash of the preceding block, altering any block could disrupt the whole chain's consistency. Additionally, blockchain operates as a disbursed ledger with multiple copies of the identical blockchain living on exceptional nodes. This redundancy further ensures that any attempts to tamper with facts are without problems detectable.

Cryptographic security: Blockchain utilizes advanced cryptographic strategies to stabilize transactions and control the right of entry. Private and public keys are at the coronary heart of this safety. Public keys, which might be derived from non-public keys, act as addresses through which customers are diagnosed at the blockchain. Private keys remain personal and are used for digital signatures to validate transactions. When consumers initiate a transaction, they signal it with their private key, developing a digital signature. This signature is precise to the transaction and the person's personal key. Upon receiving the transaction, nodes in the community confirm the virtual signature of the usage of the corresponding public key. If the signature is valid, it confirms the transaction's authenticity and integrity. Figure 2.3 illustrates the use of cryptography in blockchain through transaction security with digital signatures and key signing.

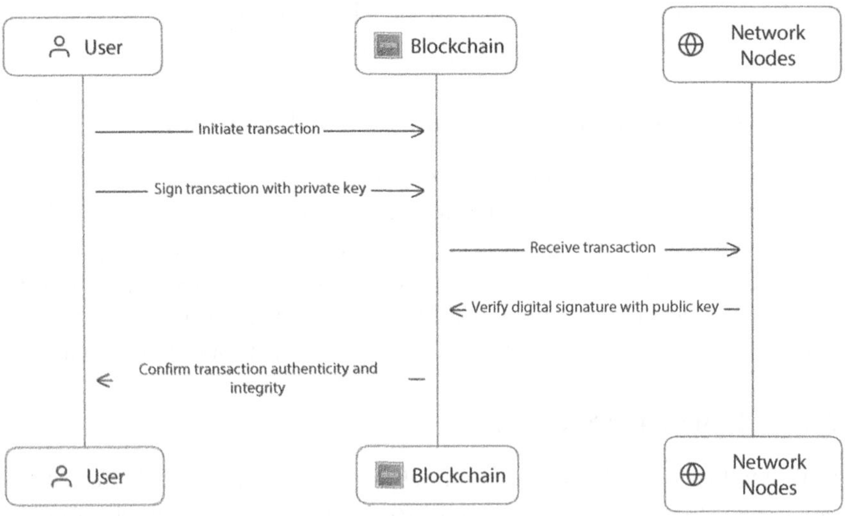

Figure 2.3 Blockchain with cryptography techniques.

Tamper resistance: Tamper resistance is a fundamental characteristic of blockchain that makes changing statistics within the blockchain extremely challenging. To tamper with an unmarried block, an attacker might need to recalculate its hash and the hashes of all next blocks. Furthermore, blockchain networks operate on the principle of consensus, wherein a majority of nodes must agree on the validity of transactions and blocks. Altering statistics requires gaining control over more than 50% of the network's computing power, a feat referred to as a 51% attack. Such attacks are not the most effective and economically feasible but also self-defeating due to the fact the attacker might efficiently break the credibility of the blockchain, undermining any ability gained from tampering. As an end result, the tamper-resistant nature of blockchain deters malicious actors and gives a remarkable degree of safety for saved records.

2.4.3 Transparency and Auditability

Transparency and auditability are critical attributes of blockchain technology that address the need for trust and accountability in various processes. These features are achieved using public ledgers and real-time updates, enhancing the reliability and integrity of blockchain-based systems.

Public ledgers: Many blockchain implementations, along with outstanding examples like Bitcoin and Ethereum, function on public ledgers. In this context, a ledger refers to a report of all transactions that have taken place at the blockchain. The term "public" signifies that this ledger is offered to absolutely everyone, allowing everybody to view the whole history of transactions. Each transaction is recorded in a block, and every block is related to the preceding one through cryptographic hashes, forming the blockchain.

This transparency has several implications:
Fraud prevention: Public ledgers serve as a deterrent against fraudulent activities. Because the transaction history is publicly available, any attempt to insert false or unauthorized transactions becomes immediately noticeable to all participants.
Independent verification: Participants in the community can independently confirm the validity of transactions with the aid of analyzing the ledger. This self-verification mechanism reduces the need for intermediaries and engenders to accept as true among contributors.
Accountability: Public ledgers foster accountability as all transactions are visible. Participants are accountable for their actions since any malicious or fraudulent behavior will be evident to the entire network.

Real-time updates: Every participant in a blockchain network has access to the identical version of the blockchain. This uniformity ensures that each individual is running with the most updated statistics. As new transactions are delivered to the blockchain, they become visible to all community participants without delay, facilitating actual-time updates.

The significance of real-time updates includes:
Reduced disputes: The synchronized nature of blockchain updates minimizes the chances of disputes arising from inconsistent or outdated data. All parties are working with the same information, reducing the risk of conflicting records.
Efficient processes: Real-time updates streamline processes that require multiple parties to access and work with the same dataset. This can be particularly advantageous in supply chain management, where stakeholders must collaborate on real-time data.
Enhanced decision-making: Accurate and current information enables better decision-making. Participants can base their decisions on the latest data, leading to more informed choices.

2.4.4 Use Cases

Cryptocurrencies: Bitcoin, the first application of blockchain, introduced the concept of digital currencies. Blockchain's security and transparency features underpin the integrity of transactions and holdings in the cryptocurrency ecosystem.
Supply chain management: Blockchain enables the traceability of products throughout the supply chain by recording each step on an immutable ledger. This increases transparency, reduces fraud, and enhances consumer trust.
Smart contracts: Self-executing smart contracts automate and implement contractual agreements, which will lower the need for intermediaries and enhance performance and transparency.
Healthcare and identity management: Blockchain can securely store clinical information and private identity information, giving people control over their facts and lowering the threat of unauthorized access.
Voting systems: Blockchain can provide a tamper-proof and transparent voting platform, increasing trust in elections and ensuring accurate results.
Figure 2.4 captures the real-time application of blockchain in the financial sector, supply chains, and smart contracts.

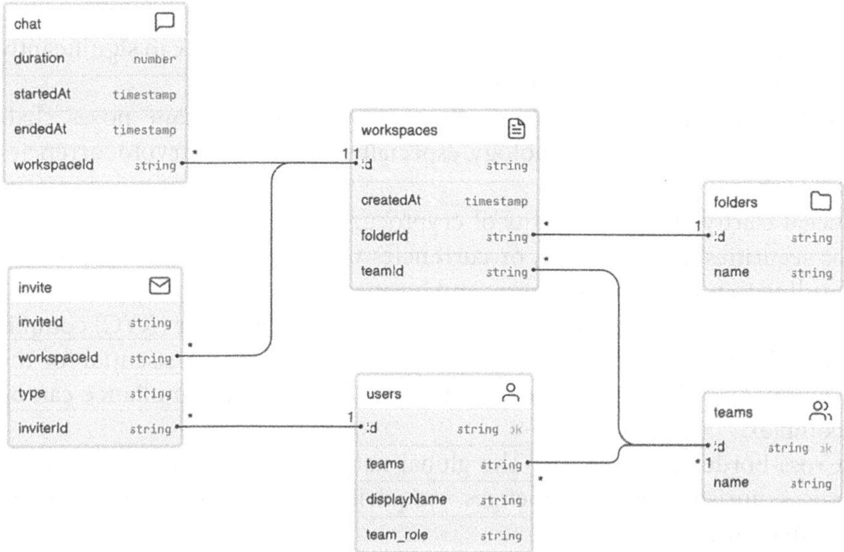

Figure 2.4 Use cases of real-time blockchain technology.

2.4.5 Challenges and Prospects

Blockchain technology gives numerous benefits. However, it also faces positive challenges that need to be addressed for its broader adoption and endured evolution. Additionally, its destiny prospects are shaped by advancements and regulatory developments.

Scalability: Scalability refers to a blockchain's capacity to handle a developing number of transactions successfully without sacrificing performance. Traditional blockchains like Bitcoin and Ethereum face scalability demanding situations due to their inherent layout.
Transaction throughput: Current blockchains have limited transaction processing capabilities. As the number of users and transactions increases, network congestion can lead to slower confirmation times and higher fees.
Ethereum's gas fees: Ethereum, a widely used blockchain platform, has faced issues with high gas fees during periods of high demand. This makes micro transactions economically unfeasible and impacts user experience.

➢ ***Solutions, including the following, are being explored:***
Layer 2 scaling: Implementations like the Lightning Network for Bitcoin and solutions like Polygon for Ethereum aim to process transactions off-chain or more efficiently.

Blockchain sharding: Sharding divides the blockchain into smaller parts, or shards, each capable of processing its transactions. This can significantly increase transaction throughput.
Regulatory uncertainty: The evolving regulatory landscape poses challenges for blockchain technology, especially in relation to cryptocurrencies and tokenized assets.
Legal clarity: The legal status of cryptocurrencies and their classification as securities, commodities, or currencies varies from country to country, leading to uncertainty for users and businesses.
Anti-money laundering (AML) and Know-Your-Customer (KYC) compliance: Regulations related to AML and KYC practices are essential for the legitimacy of blockchain-based financial services, but compliance can be complex.
Cross-border transactions: The global nature of blockchain transactions raises jurisdictional challenges, requiring harmonized international regulations.
Environmental concerns: Blockchain's environmental impact, particularly for PoW blockchains like Bitcoin, has raised concerns due to the energy-intensive mining process.
Energy consumption: PoW mining requires powerful computational resources and consumes significant electricity, leading to debates about its sustainability.
Carbon footprint: Large-scale PoW mining operations can contribute to carbon emissions and environmental degradation.

➢ ***Efforts to address these concerns include:***
Proof of Stake: PoS mechanisms, adopted by networks like Ethereum 2.0, require validators to hold and lock a certain amount of cryptocurrency rather than solving complex mathematical puzzles. This reduces energy consumption.
Hybrid approaches: Some projects explore hybrid consensus mechanisms that combine PoW and PoS to strike a balance between security and energy efficiency.

2.4.6 Synergy Between Quantum Cryptography and Blockchain

The integration of QC and blockchain technology represents an effective fusion that addresses some of the essential demanding situations confronted by way of each domain name. While QC gives extraordinary stages of security, blockchain gives a decentralized and tamper-resistant infrastructure. This aggregation can reshape the landscape of data safety, privacy, and consensus mechanisms.

Quantum-resistant cryptography: Traditional cryptographic techniques, which rely on mathematical algorithms, are liable to assaults from quantum computer systems due to their potential to solve complex troubles speedy. QC introduces a brand-new paradigm wherein cryptographic protocols harness the standards of quantum mechanics to create inherently steady verbal exchange channels. This quantum-resistant cryptography addresses the ability posed by quantum computers to classical cryptographic structures.
Enhanced data security: By using QC-based encryption and key distribution, blockchain networks can gain a brand-new degree of security. Quantum Key Distribution (QKD) guarantees that cryptographic keys are generated and exchanged with a remarkable degree of protection, as any strive at interception might disrupt the quantum states, consequently revealing the presence of eavesdropping. This enhances the confidentiality and integrity of transactions and statistics stored on the blockchain.
Quantum-proof consensus mechanisms: Quantum computing has the potential to undermine traditional consensus mechanisms, inclusive of PoW and PoS, which steady blockchain networks. However, quantum-proof consensus algorithms are being advanced to face up to assaults from each classical and quantum computer. These mechanisms make sure the continued security of blockchain networks even inside the technology of quantum computing.
Decentralized trust and quantum randomness: Blockchain's decentralized nature aligns well with the principles of QC. Quantum Random Number Generators (QRNGs) can provide true randomness that is difficult to predict, enhancing the security of various processes within blockchain networks. These QRNGs can be used for creating unpredictable seeds for cryptographic operations, enhancing the resilience of blockchain systems against attacks.
Tokenization of quantum data: As QC evolves, it could generate vast amounts of valuable information. Blockchain can provide a stable and transparent platform for tokenizing and managing this data, allowing steady sharing, monetization, and traceability of quantum records belongings.
Quantum auditing and proofs: Quantum properties, such as entanglement, can be leveraged to provide irrefutable evidence of certain states. This can be integrated into blockchain systems to create quantum audit trails, allowing for provably secure tracking of transactions and data changes. Figure 2.5 captures the convergence of the two worlds of quantum cryptography and blockchain demonstrating quantum key distribution, trustless systems, consensus, and voting frameworks.

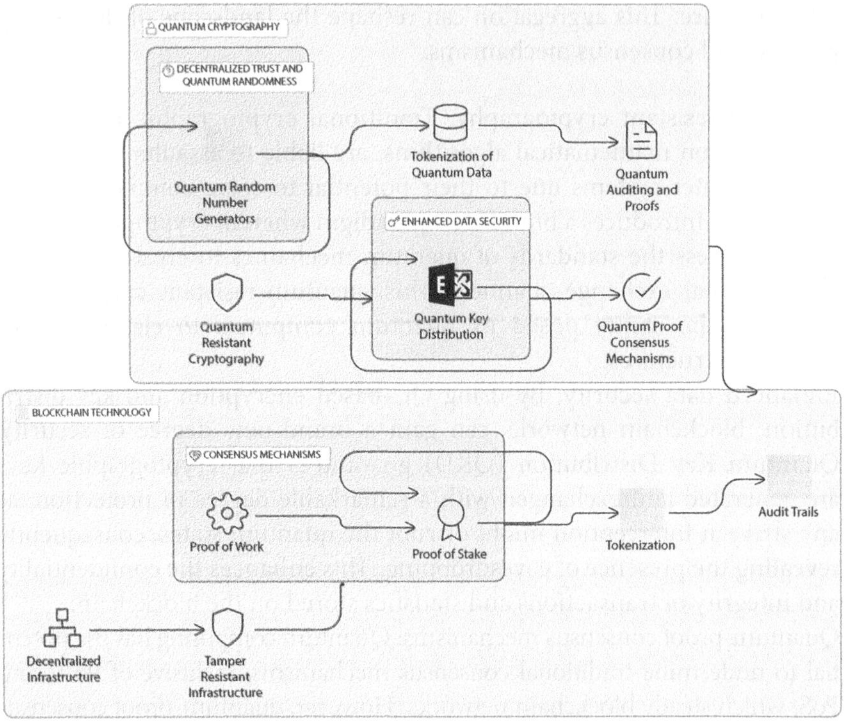

Figure 2.5 Quantum cryptography and blockchain technology.

2.5 Flying *Ad Hoc* Networks: A Dynamic Communication Infrastructure

Flying *ad hoc* networks represent a specialized shape of mobile *ad hoc* networks (MANETs), wherein unmanned aerial vehicles (UAVs), generally referred to as drones, serve as the nodes for verbal exchange [20]. The FANETs introduce a dynamic and versatile communication infrastructure that has gained giant interest due to their capacity applications in diverse fields, together with disaster control, surveillance, environmental monitoring, and communication in far-off or hard terrains [16].

2.5.1 Key Characteristics and Components

Autonomous UAVs: FANETs are composed of UAVs equipped with communication modules, sensors, and navigation systems. These UAVs operate autonomously, making decisions based on board algorithms, mission objectives, and real-time environmental inputs. Figure 2.6 describes the

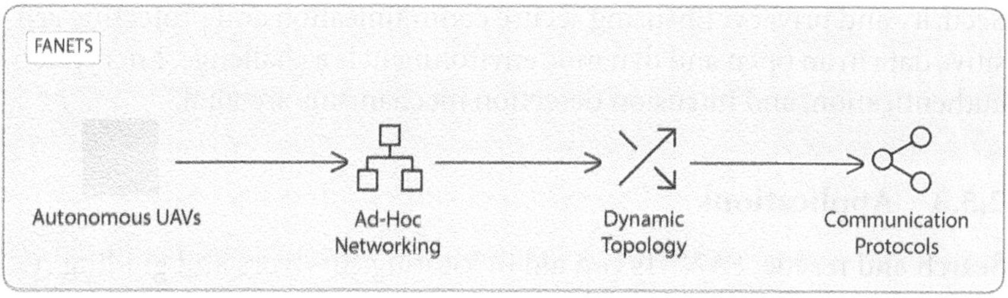

Figure 2.6 Components of FANETs.

structure of FANETs, focused on autonomous UAV and their constituents: *Ad-hoc* network, dynamic topology, and communicating protocols.

Ad hoc networking: Unlike traditional networks with fixed infrastructure, FANETs employ *ad-hoc* networking, enabling UAVs to establish communication links on the fly without relying on pre-existing infrastructure. The UAVs collaborate to form a temporary network for data exchange.
Dynamic topology: The dynamic nature of the UAVs' mobility leads to constantly changing network topologies. The UAVs can move in any route, and they can be a part of or go away from the network at any time. This dynamic topology requires adaptive and resilient communique protocols.
Communication protocols: FANETs require specialized communication protocols that can handle the UAVs' mobility, intermittent connectivity, and rapidly changing network conditions. These protocols include routing, data dissemination, and link management strategies Amponis *et al.* (2021) [19].

2.5.2 Challenges and Considerations

Connectivity maintenance: Maintaining reliable communication links in a dynamic aerial environment is a significant challenge. The UAVs must adapt to varying distances, altitudes, and potential obstacles [21].
Interference and spectrum management: With multiple UAVs communicating simultaneously, interference can degrade signal quality. Efficient spectrum management and interference mitigation strategies are essential.
Energy efficiency: UAVs are often powered by batteries with limited capacities. Efficient energy management and communication strategies are crucial to maximizing flight duration.
Mobility management: UAV mobility affects signal strength and link quality. Mobility-aware protocols must handle fast-moving nodes and ensure seamless handovers.

Security and privacy: Ensuring secure communication and protecting sensitive data in an open and dynamic environment is a challenge. Encryption, authentication, and intrusion detection mechanisms are vital.

2.5.3 Applications

Search and rescue: FANETs can aid in locating survivors and guiding rescue operations in disaster-stricken areas where traditional communication networks are unavailable or damaged.
Surveillance and monitoring: UAVs in FANETs are well-suited for large-area surveillance, such as monitoring traffic, tracking wildlife, or detecting environmental changes.
Communication infrastructure: FANETs can establish temporary communication infrastructure in remote or isolated areas, providing emergency communication channels.
Precision agriculture: UAVs can be deployed for crop monitoring, data collection, and spraying pesticides or fertilizers in precision agriculture applications [17].

2.5.4 Convergence of Quantum Cryptography, Blockchain, and FANETs

The convergence of QC, blockchain technology, and FANETs represents a transformative synergy that can address critical challenges in secure communication, data integrity, and dynamic networking. This integration leverages the unique capabilities of each component to create a comprehensive communication ecosystem with enhanced security, privacy, and adaptability [18].

Quantum cryptography and blockchain: QC offers unbreakable encryption and secure key distribution, solving a fundamental challenge in secure communication. By integrating QC with blockchain technology, the following enhancements can be achieved:
Quantum-safe key management: QC provides an ultra-secure method for generating and distributing cryptographic keys. Blockchain's tamper-resistant nature ensures the integrity of key exchange, reducing the risk of key compromise.
Immutable records: Blockchain's immutability complements QC's security by creating an auditable record of key exchange and transactions. This prevents unauthorized modifications and provides an accountable trail.

Decentralized trust: The decentralized nature of blockchain aligns with QC's goals of secure communication without relying on a central authority. This combination ensures trust and security without intermediaries.

Blockchain and FANETs: Integrating blockchain with FANETs introduces transparency, security, and coordination to dynamic drone communication networks.

Decentralized flight data: Blockchain can securely store flight data, including UAV trajectories, maintenance records, and communication logs. This ensures transparency and accountability in FANET operations.

Secure data sharing: Blockchain's permissioned access allows UAVs to securely share critical data such as mission updates, weather conditions, and sensor readings among network participants.

Smart contracts for coordination: Smart contracts can automate tasks like routing optimization, resource allocation, and dynamic reconfiguration of FANETs, enhancing coordination among drones. Figure 2.7 depicts the trio of quantum cryptography, blockchain, and FANETs, accentuating their importance in secure communication and securing the integrity of the data.

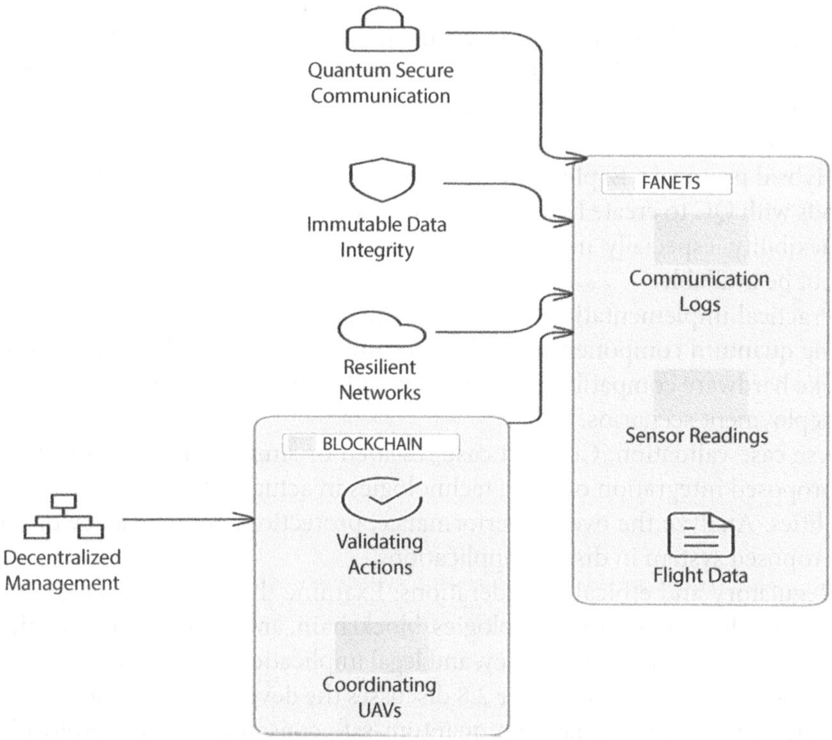

Figure 2.7 Integration of quantum cryptography, blockchain, and FANETs.

Quantum cryptography, blockchain, and FANETs integration: The combined use of QC, blockchain, and FANETs leads to a holistic communication ecosystem.
Quantum-secure communication: QC ensures unbreakable encryption and key distribution, safeguarding data in transit across FANETs.
Immutable data integrity: Blockchain guarantees the integrity of data stored in FANETs, whether it is flight data, sensor readings, or communication logs.
Decentralized management: Blockchain's decentralized structure complements FANETs by providing a secure platform for coordinating UAVs and validating their actions.
Resilient networks: FANETs' dynamic topology aligns with blockchain's adaptability, allowing the network to adjust to UAV mobility and changing conditions.

2.6 Future Directions and Challenges

2.6.1 Future Directions

Quantum-safe blockchain consensus: Investigate how quantum computers could potentially break current blockchain consensus mechanisms. Explore the improvement of quantum-resistant consensus algorithms to ensure blockchain safety inside the quantum era.
Hybrid protocols: Explore the integration of classical cryptographic methods with QC to create hybrid protocols that provide enhanced security and flexibility, especially in scenarios where full quantum capabilities might not be available.
Practical implementation: Investigate the practical challenges of integrating quantum components with blockchain and FANETs. Consider aspects like hardware compatibility, scalability, and energy efficiency in real-world deployment scenarios.
Use case validation: Conduct case research or simulations to validate the proposed integration of those technologies in actual-international eventualities. Analyze the overall performance, protection, and efficiency of the proposed system in distinct applications.
Regulatory and ethical considerations: Examine the regulatory landscape surrounding quantum technologies, blockchain, and drones. Address ethical considerations, data privacy, and legal implications in the context of the proposed convergence. Figure 2.8 discusses the development of blockchain technology with emphasis on quantum-safe consensus, hybrid protocols, and legal boundaries.

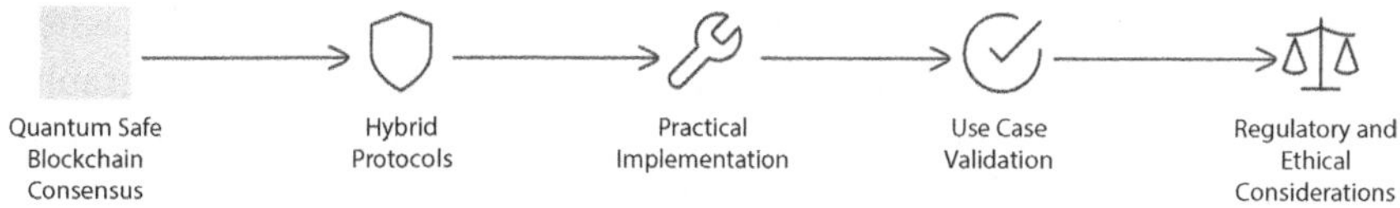

Figure 2.8 Future implementations of blockchain technology.

2.6.2 Challenges

Technical integration complexity: Integrating QC, blockchain, and FANETs requires dealing with the complexity of each technology and ensuring seamless interoperability.
Quantum hardware limitations: The availability and reliability of quantum hardware pose challenges. Investigate how to work within the constraints of current quantum technologies while considering future advancements.
Interdisciplinary expertise: Research at the intersection of quantum physics, cryptography, blockchain, and UAV technologies demands a multidisciplinary approach. Collaborating with experts from various fields is essential.
Performance trade-offs: Assess potential trade-offs in terms of latency, computational overhead, and energy consumption when integrating these technologies, especially in resource-constrained environments like FANETs.
Validation and real-world testing: Experimentation and validation in actual global settings can be hard because of the dynamic nature of FANETs and the restrictions of to-be-had quantum hardware.
Quantum cryptanalysis: Address the possibility of quantum computers breaking current quantum cryptographic techniques and the impact this might have on the proposed convergence.
Standardization: The lack of standardized protocols for integrating quantum and blockchain technologies could pose challenges for interoperability and adoption.

2.7 Conclusion

In conclusion, our comprehensive chapter embarked on a journey to unravel the intricate tapestry woven by the convergence of QC, blockchain technology, and FANETs. Through an exhaustive survey backed by mathematical analysis, we have unveiled a landscape of possibilities that stands to reshape the paradigms of secure communication, data integrity, and dynamic networking. Our exploration of QC illuminated a realm where the principles of quantum mechanics intersect with the cryptographic domain. With an emphasis on unbreakable key exchange and quantum-resistant cryptography, we probed the mathematical foundations that ensure secure communications against the looming quantum threat. The analysis of these cryptographic primitives has underscored the ironclad security they offer, fortified by the unassailable principles of quantum physics.

References

1. Hafeez, S., Khan, A.R., Al-Quraan, M., Mohjazi, L., Zoha, A., Imran, M.A., Sun, Y., Blockchain-Assisted UAV Communication Systems: A Comprehensive Survey. *IEEE Open J. Veh. Technol.*, 4, 558–580, 2023, DOI: 10.1109/OJVT.2023.3295208.
2. Hassija, V., *et al.*, Fast, Reliable, and Secure Drone Communication: A Comprehensive Survey. *IEEE Commun. Surv. Tutor.*, 23, 4, 2802–2832, Fourthquarter 2021, doi: 10.1109/COMST.2021.3097916.
3. Wu, M., Wang, K., Cai, X., Guo, S., Guo, M., Rong, C., A comprehensive survey of blockchain: From theory to IoT applications and beyond. *IEEE Internet Things J.*, 6, 5, 8114–8154, 2019.
4. Abdulhae, O.T., Mandeep, J.S., Islam, M., Cluster-Based Routing Protocols for Flying *Ad Hoc* Networks (FANETs). *IEEE Access*, 10, 32981–33004, 2022, doi: 10.1109/ACCESS.2022.3161446.
5. Ali, H., Islam, S.U., Song, H., Munir, K., A performance-aware routing mechanism for flying *ad hoc* networks. *Trans. Emerging Telecommun. Technol.*, 32, 1, e4192, 2021.
6. Rovira-Sugranes, A., Razi, A., Afghah, F., Chakareski, J., A review of AI-enabled routing protocols for UAV networks: Trends, challenges, and future outlook. *Ad Hoc Netw.*, 130, 102790, 2022.
7. Khan, I.U., Qureshi, I.M., Aziz, M.A., Cheema, T.A., Shah, S.B.H., Smart IoT control-based nature inspired energy efficient routing protocol for flying *ad hoc* network (FANET). *IEEE Access*, 8, 56371–56378, 2020.
8. Pasandideh, F., da Costa, J.P.J., Kunst, R., Hardjawana, W., de Freitas, E.P., A systematic literature review of flying ad hoc networks: State-of-the-art, challenges, and perspectives. *J. Field Rob.*, 40, 4, 955–979, 2023.
9. Kumar, A., Bhatia, S., Kaushik, K., Manjula Gandhi, S., Gayathri Devi, S., Pacheco, D.A.J., Mashat, A., Survey of promising technologies for quantum drones and networks. *IEEE Access*, 9, 125868–125911, 2021.
10. Liu, H.-Y., Tian, X.-H., Gu, C., Fan, P., Ni, X., Yang, R., Zhang, J.-N., *et al.*, Drone-based entanglement distribution towards mobile quantum networks. *Natl. Sci. Rev.*, 7, 5, 921–928, 2020.

11. Fernandez-Carames, T.M. and Fraga-Lamas, P., Towards post-quantum blockchain: A review on blockchain cryptography resistant to quantum computing attacks. *IEEE Access*, 8, 21091–21116, 2020.
12. Wang, L., Shen, X., Li, J., Shao, J., Yang, Y., Cryptographic primitives in blockchains. *J. Netw. Comput. Appl.*, 127, 43–58, 2019.
13. Longo, R., Mascia, C., Meneghetti, A., Santilli, G., Tognolini, G., Adaptable Cryptographic Primitives in Blockchains *via* Smart Contracts. *Cryptography*, 6, 3, 32, 2022.
14. Bhatt, A.P. and Sharma, A., Quantum cryptography for internet of things security. *J. Electron. Sci. Technol.*, 17, 3, 213–220, 2019.
15. Islam, A., Rahim, T., Masuduzzaman, M.D., Shin, S.Y., A blockchain-based artificial intelligence-empowered contagious pandemic situation supervision scheme using internet of drone things. *IEEE Wireless Commun.*, 28, 4, 166–173, 2021.
16. Yazid, Y., Ez-Zazi, I., Guerrero-Gonzalez, A., El Oualkadi, A., Arioua, M., UAV-enabled mobile edge-computing for IoT based on AI: A comprehensive review. *Drones*, 5, 4, 148, 2021.
17. Vashishth, T.K., *et al.*, Advanced Technologies and AI-Enabled IoT Applications in High-Tech Agriculture, in: *Handbook of Research on AI-Equipped IoT Applications in High-Tech Agriculture*, A. Khang (Ed.), pp. 155–166, IGI Global, Hershey, PA, 2023.
18. Alzoubi, Y.I., Al-Ahmad, A., Kahtan, H., Jaradat, A., Internet of things and blockchain integration: security, privacy, technical, and design challenges. *Future Internet*, 14, 7, 216, 2022.
19. Amponis, G., Lagkas, T., Sarigiannidis, P., Vitsas, V., Fouliras, P., Wan, S., A survey on FANET routing from a cross-layer design perspective. *J. Syst. Archit.*, 120, 102281, 2021.
20. Gupta, M., Vashishth, T.K., Verma, P.K., Machine learning and deep learning based intrusion detection for blackhole attacks in mobile *ad-hoc* networks. *Multi. Sci. J.*, 6, 11, 2024209–2024209, 2024.
21. Ahamed Ahanger, T., Aldaej, A., Atiquzzaman, M., Ullah, I., Yousufudin, M., Distributed blockchain-based platform for unmanned aerial vehicles. *Comput. Intell. Neurosci.*, 2022, 1–16, Aug. 2022. doi: 10.1155/2022/4723124.

3

A Survey on Security Issues, Challenges, and Future Perspectives on FANETs

Syed Mohd Faisal[1]*, Wasim Khan[1], Mohammad Ishrat[1] and Taskeen Zaidi[2]

[1]Department of CSE, Koneru Lakshmaiah Education Foundation, Vaddeswaram, Andhra Pradesh, India
[2]Department of CS&IT, Jain (Deemed-to-be-University), Bengaluru, India

Abstract

The efficiency of unmanned aerial vehicles (UAV) has improved recently due to advances in digital and electronics systems, particularly the shrinking and cost reduction of devices. A flying *ad hoc* network (FANET), which is a part of a mobile *ad hoc* network (MANET), is the name of the UAV network. The UAVs are currently highly common in the field of civil use in addition to the military. Compared to a lone UAV system, a group of UAVs can work together to accomplish missions more affordably and effectively. As a result, this decision sparked the creation of fresh networking techniques for UAVs, as well as ground control stations. Before using FANETs effectively to deliver dependable and robust context-specific networks, there are a number of issues that must be solved. In this chapter, we aim to deliver a thorough assessment of the current communication infrastructures, security issues, and future perspectives. We discuss the security concerns that must be overcome, explain outstanding issues with FANET networking, and reveal the routing protocols, security requirements, mobility, security issues, and trajectory optimization models utilized in a FANET to address UAV communication and collaboration concerns. Our objective is to give researchers a general understanding of the various issues that need to be tackled in this field.

***Keywords*:** UAV, FANET, MANET, IDS, cryptography, routing, drone, attack

**Corresponding author*: smfaisalcse@gmail.com

Taskeen Zaidi, Adarsh Kumar, Saurav Mallik and Keshav Kaushik (eds.) Autonomous Flying Ad-Hoc Networks, (37–86) © 2025 Scrivener Publishing LLC

3.1 Introduction

Over the past few years, the global market for unmanned aerial vehicles (UAVs) achieved humongous growth and received additional spotlight from governments and corporations. This is because UAVs have many civilian and military uses, including traffic monitoring, search operation, rescue operation, oversight, and bio-chemical sensing. The use of UAVs for passenger transportation—also known as "air taxis"—is now the subject of a socio-technical dispute. These vehicles can perform electric vertical take-off and landing operations, which will eventually replace commercial helicopters. The business drone market is expected to bring in 129.3 billion dollars by 2025, according to a recent study [1]. The United States Federal Aviation Administration (FAA) says that the business of drone markets could be three times bigger than it is by the end of 2023. So, putting UAVs on the public market will make more people want to use them in business in many different areas. Due to the growth of the drone market, big multinational corporations and businesses are now interested in investing in UAVs. So, UAVs will be an important part of industrial society because more civilians are using them or are ready to use them. Figure 3.1 illustrates a broader perspective of the FANET, highlighting its distinctions from MANET and VANET.

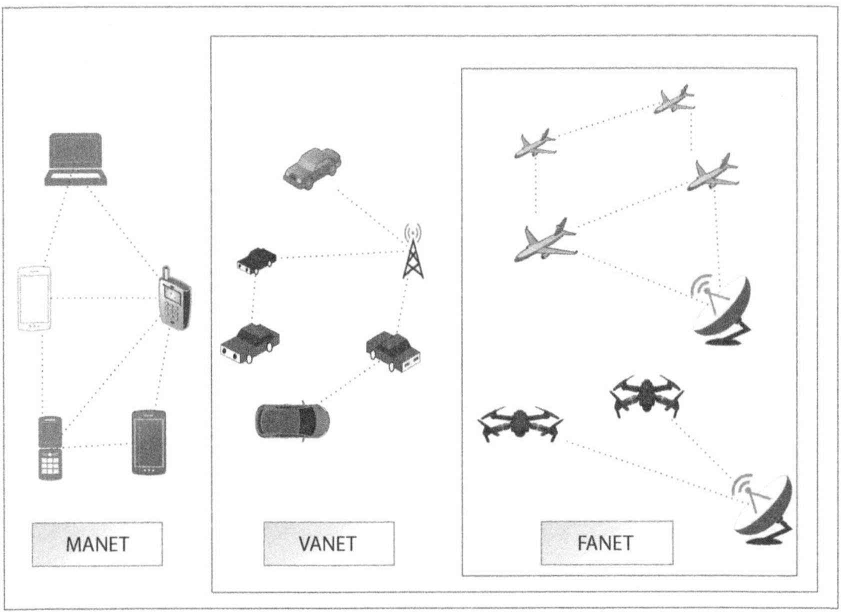

Figure 3.1 Comparison of MANET, VANET, and FANET architectures with representative node types.

Even though the drone industry is growing around the world, and industrial UAVs have a lot of benefits, there are a lot of drone mishaps recorded every day. So, we need to come up with good ways to stop these kinds of things from happening. To do this, one line of thought says that UAV hazards should be found and named as soon as possible. This method would give the operator enough time to put in place the tools needed to stop such hazards and threats. It is very important to think about how this technology could be used maliciously and how it could harm people who use it for a good cause. In fact, the growth of UAVs at an exponential rate makes their software and hardware parts more vulnerable [2]. As the number of UAV threats grows, security and privacy issues related to putting UAVs in the aerospace industry have become a big concern for governments. Additionally, most commercially available UAVs lack fundamental security topographies such as intrusion detection systems (IDS). Therefore, the attackers' UAVs are soft target to attack or gain access.

There have been a number of successful cyberattacks recently against UAVs [3]. Cyberattacks on private UAVs in t real world are now considered a matter of national security. There has been debates in the public and corporate sectors about security concerns brought up by the introduction of UAVs into the national airspace. From a security perspective, there are many kinds of cyber threats, which show that UAVs are exposed in different ways. Indeed, malicious actuators get an advantage due to the widespread use of drones in legitimate industries. They take advantage of different vulnerabilities in public and private drones to put people's safety in danger. Also, drone makers do not think about privacy and security issues early in the process of making them. In addition, drone manufacturers do not give serious consideration to security and privacy issues until late in the development process.

The dynamic usage of public UAVs in various classes might raise new privacy and security concerns, and this is something that must be mentioned. Existing approaches to identify hacked drones and safeguard drone systems are, therefore, inadequate. In this way, cyber strikes against UAVs are possible because the right security measures have not been put in place to guarantee the confidentiality, integrity, and availability of the system. Thus, it is crucial to examine UAVs from a privacy and safety standpoint. Contrary to what we have discussed, putting UAVs in the sky can correspondingly invade the privacy of individuals and cause problems in sensitive places like chemical plants and nuclear plants. In fact, most UAVs have cameras on board, which could possibly reveal private information about what people do.

In general, we think of UAVs as complicated aircraft. A flying UAV has onboard sensors, like an accelerometer and global positioning system (GPS), that send sensor data to the flight controller then the flight controller forwards the data to the operator through a defined route. The operator processes the incoming signals and sends the received signal to the flight supervisor/controller. Four primary parts must work in tandem for the UAV system to remain in its ideal configuration. These parts are the hardware, the program, the data link, and the sensors. Additionally, the UAV system may crash and burn if any of its parts fail or start working maliciously.

3.2 Architecture of FANET

In this section, we see the components of UAVs and how they are incorporated with FANETs. Later, we see the communication architecture of UAVs. The term internet of drones (IoD) is becoming popular nowadays, which has proved services to the drone in the air space. Using FANET zones, the IoD may build a network of several drones that communicate with one another organizationally without any human intervention. The general architecture of UAVs includes an unmanned aircraft, ground control station (GCS), which includes devices like laptops, computers, mobile devices, and tablets that are used to pilot UAVs, and the last component of architecture is the communication link, which is used for communication among GCS and UAV. Basically, there were two forms of communication possible: control communication and data communication (Figures 3.2–3.9). In control communication, the GSC sends the pilot commands to the UAVs. In data communication, the UAVs forward signals and capture data to the GCS.

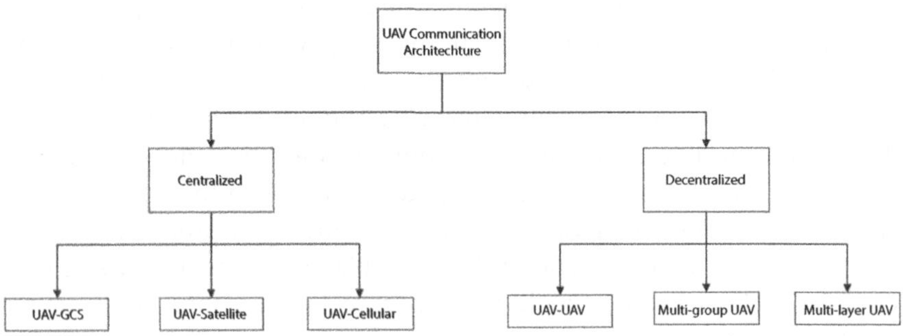

Figure 3.2 Types of communication.

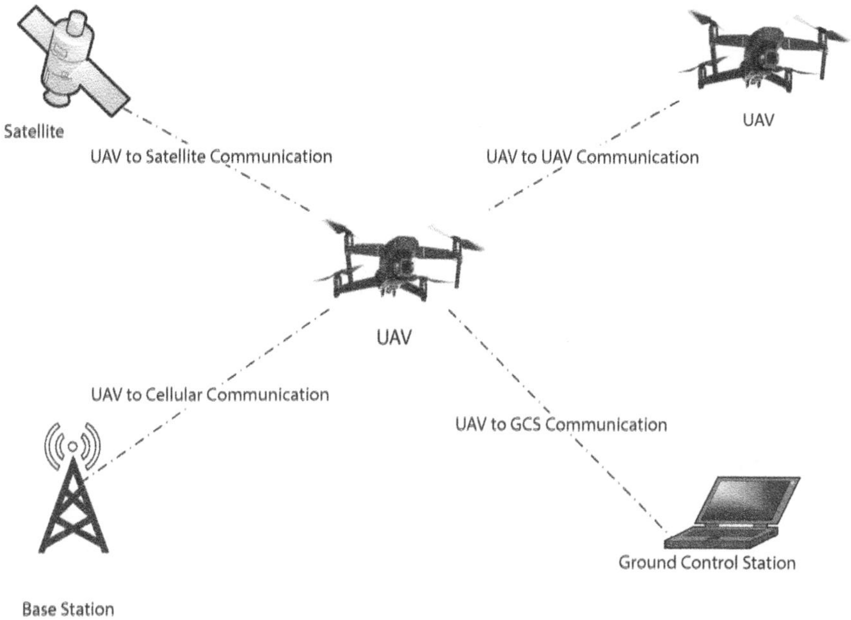

Figure 3.3 Broad view of communication.

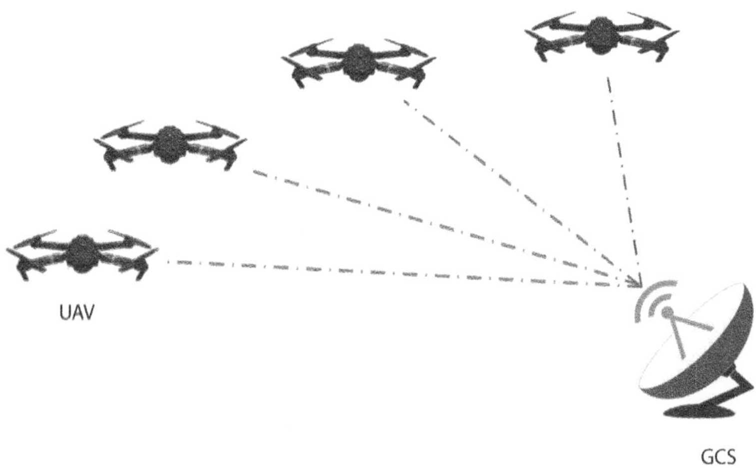

Figure 3.4 Centralize UAV GCS communication.

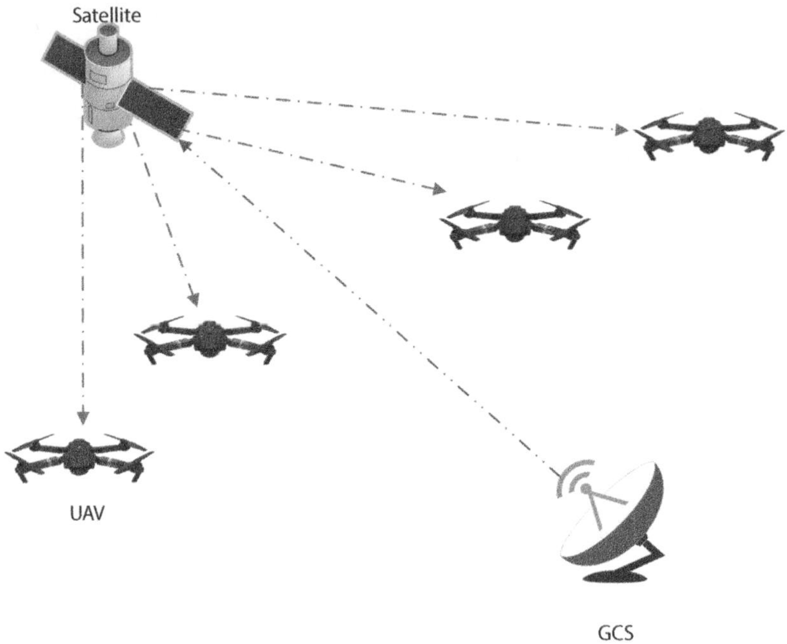

Figure 3.5 Centralize UAVs satellite communication.

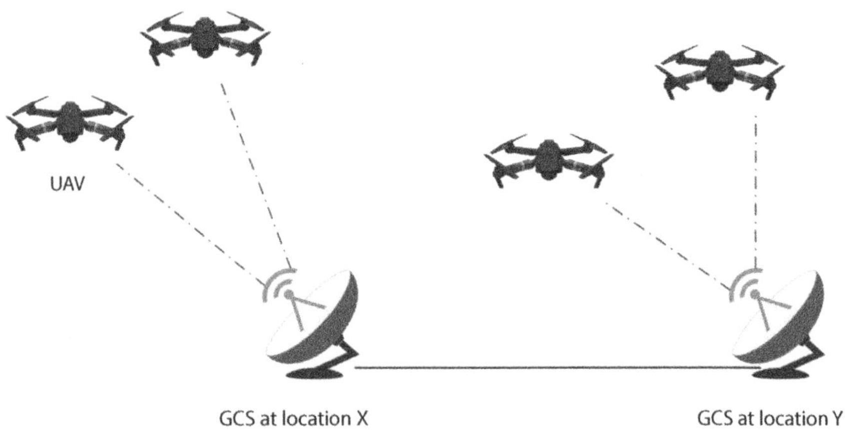

Figure 3.6 Centralize UAVs satellite communication.

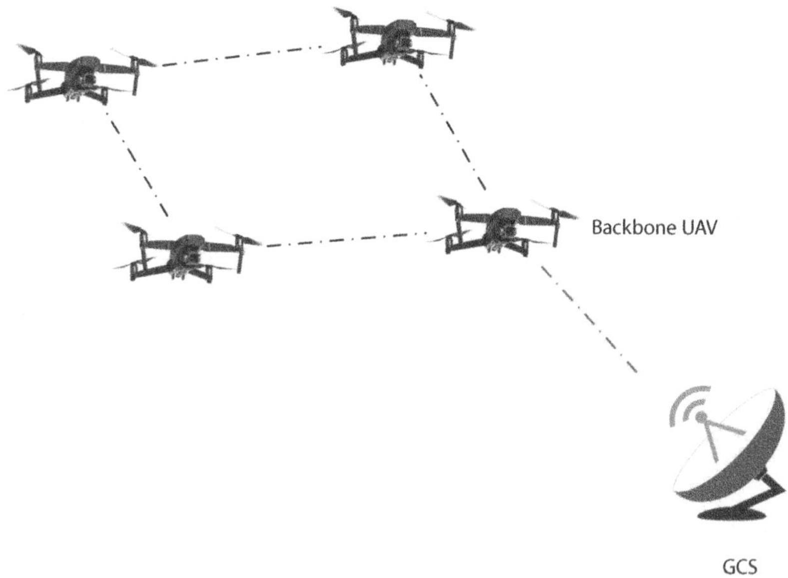

Figure 3.7 Decentralize a UAV *ad hoc* network.

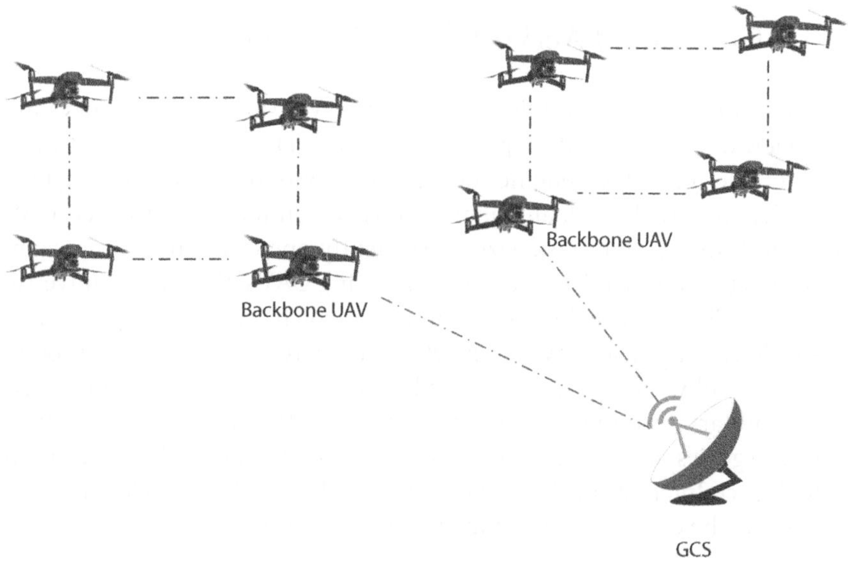

Figure 3.8 Decentralize multigroup UAVs *ad hoc* network.

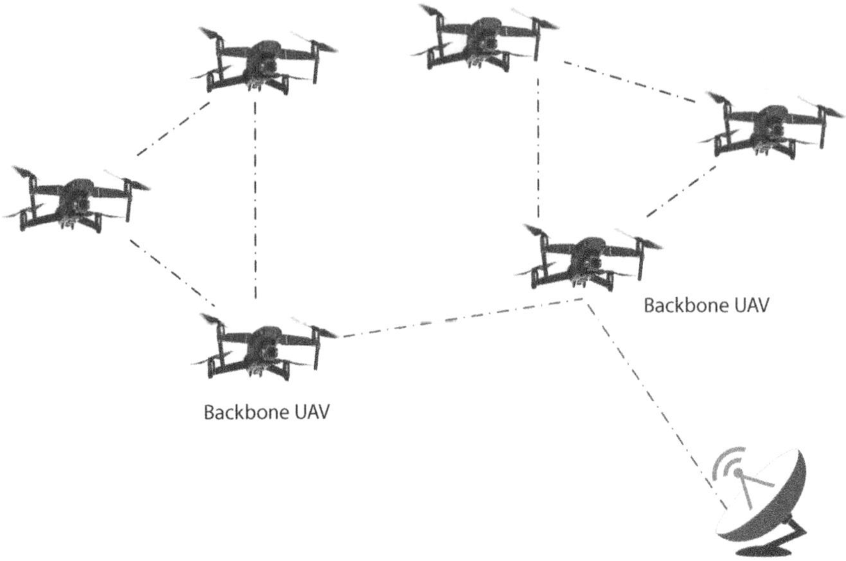

Figure 3.9 Decentralize multilayer UAVs *ad hoc* network.

3.3 Unmanned Aerial Vehicle Classification

In the section, we classify different types of UAVs. This literature gives a full overview of UAVs considering the size (Table 3.1), weight, range, altitude, type (Table 3.2), class, engine, degree of autonomy, ownership, payload classification (Table 3.3), and applications. When identifying which UAV suits best for a certain task, size is frequently one of the most important considerations. When it comes to categorizing UAVs, there is no universal standard. Civilians use their own standards, which are constantly evolving, while defense agencies have other sets of standards for the classification of UAVs. Broadly, UAVs are categorized by size, payload, range, endurance, and a military-inspired tier system is also used. The global UAV classification categories are well-explained on the United State National Aviation Intelligence Integration (US-NAII) Office website [4]. The following subclasses can be created for classification based on size:

Table 3.1 Classification of UAVs according to their sizes.

Type	Length	Propeller diameter	Weight	Use
Very small UAVs	150 mm	Up to 51 mm	200 gm	❖ Military surveillance
Small UAVs	300 mm	76 to 152 mm	200 to 1000 gm	❖ Audits of indoor equipment ❖ Photography
Medium UAVs	300-1200mm	150 to 640 mm	1 to 20 kg	❖ Amateur photography ❖ Biological warfare
Large UAVs	120cm	64 cm	More than 20 kg	❖ Gathering meteorological data ❖ Surveillance

Table 3.2 Unmanned aerial vehicle classifications based on wing types.

Drone type	Advantage	Disadvantage	Usage
Multi-rotor drones	❖ Easy to handle and direct ❖ Portable ❖ Low Price	❖ Unstable ❖ Low speed ❖ Limited flight time ❖ Very low payload capability	❖ Photography ❖ Inspection ❖ Surveillance ❖ Agriculture ❖ Aerial inspection
Fixed-wing drones	❖ Better stability ❖ Better payload capacity ❖ Better flight time	❖ No vertical takeoff and landing ❖ Require training ❖ Plane surface is required for landing ❖ Expansive	❖ Sky-based surveying ❖ Inspection for the verification of services (i.e. power line and pipeline inspection) ❖ Surveillance ❖ Agriculture

(Continued)

Table 3.2 Unmanned aerial vehicle classifications based on wing types. (*Continued*)

Drone type	Advantage	Disadvantage	Usage
Single-rotor helicopter drones	❖ Vertical takeoff and landing ❖ Heavier payload capacity ❖ Long flight time	❖ Require expert training for handling the drone ❖ Expensive	❖ Laser scanning from above using light detection and ranging (LiDAR) ❖ Used in military
Fixed-wing hybrid VTOL drones	❖ Vertical takeoff and landing ❖ Heavier payload capacity ❖ Fast speed	❖ Require expert training for handling the drone ❖ Very expensive	❖ Search and rescue ❖ Surveillance ❖ Deep down inspection ❖ Used in the military ❖ Package delivery

3.3.1 Classification of UAVs According to the Size

3.3.1.1 Very Small UAVs

Very small UAV classifications include UAVs whose sizes are like a big bug's (i.e. between 30 to 50 cm long). A common micro-design is an insect-like UAV with rotating or flapping wings. They can be utilized for biological

Table 3.3 Drone classifications based on the payload.

Drone type	Weight	Payload capacity	Usage
Featherweight drones	Under 11gm	4 – 100gm	❖ Surveillance ❖ Intelligence gathering
Lightweight drones	200 – 1000gm	150 – 270gm	❖ Surveillance ❖ Photography
Middleweight drones	1 – 600 kg	400 – 1460 gm	❖ Aerial photography ❖ Logistics
Middleweight drones	Over 600kg	More than 1000 kg	❖ Enemy detection ❖ Film making ❖ Logistics

warfare and spying since they are so tiny and light. The standard airplane configuration is used by larger ones. The required degree of mobility determines whether to use rotating or flapping wings. Designs with flapping wings enable establishing and landing on tiny areas. The Israeli IAI Malat Mosquito

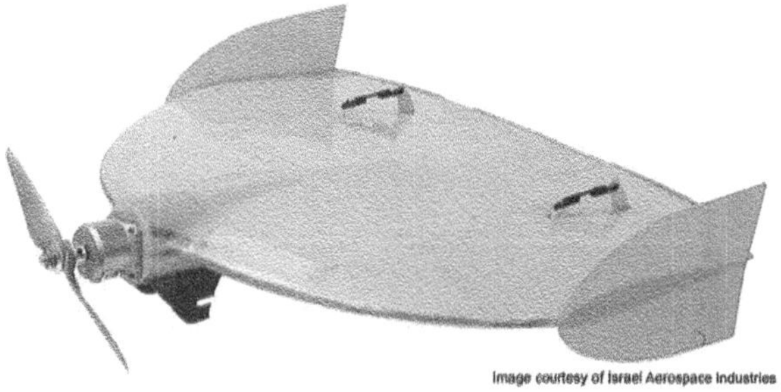

Figure 3.10 Israeli IAI malat mosquito [7–9].

Figure 3.11 The US aurora flight sciences skate [7–9].

Figure 3.12 The CyberQuad mini [7–9].

Figure 3.13 CyberQuard maxi [7–9].

Figure 3.14 Small UAV thrown by hand [7–9].

(Figure 3.10), which has a wing size of 35 cm and a 40-minute hover, the US Aurora Flight Sciences Skate (Figure 3.11), which has a wing size of 60 cm and 33 cm in length, the Australian Cyber Technology "CyberQuad Mini" (Figure 3.12), which has a 42x42 cm square, and their most recent model, the "CyberQuad Maxi" (Figure 3.13), are examples of very small UAVs [4, 7–9].

3.3.1.2 Small UAVs

The small UAVs, commonly stated as mini-UAVs, are demonstrated as UAVs with dimensions larger than 50 cm and smaller than 2 m. The fixed-wing model serves as the foundation for the design of this category, and

Figure 3.15 Turkish bayraktar [7–9].

Figure 3.16 RQ-11 raven [7–9].

the majority of them are launched by hand by being thrown into the air, as depicted in Figure 3.14. Members of this particular UAV subclass are, the Turkish Bayraktar (Figure 3.15), whose weight is around 5 kg and covers a 20 km range; the 1-meter-long RQ-11 Raven (Figure 3.16) by US AeroVironment having 1.4-meter wingspan; the American Aerospace RS-16, seen in Figure 3.17, and the US Army RQ-7 Shadow shown in Figure 3.16 [5].

The RS-16 (Figure 3.17) is a hybrid small-to-medium-sized UAV. This class of UAV includes those with rotary-wing designs [7–9].

Figure 3.17 RS16 [7–9].

Figure 3.18 RQ-5A [7–9].

Figure 3.19 RQ-2A pioneer [7–9].

Figure 3.20 UK watchkeeper [7–9].

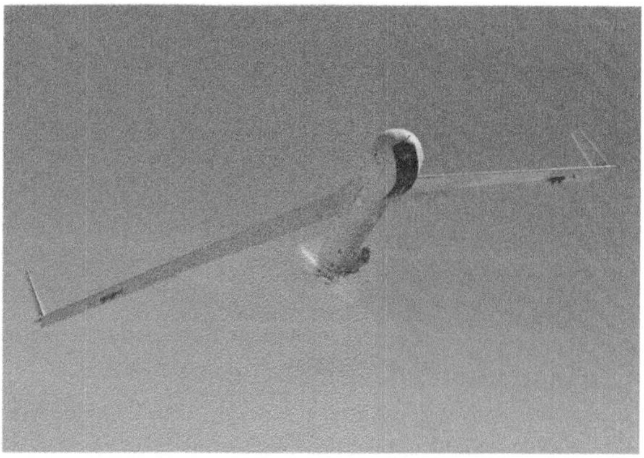

Figure 3.21 US boeing eagle eye [7–9].

3.3.1.3 Medium UAV

Medium-class UAVs are those that are bigger than small UAVs and are too bulky for a person to carry or transport. They typically have a wing size of 5 to 10 m and a payload capacity (loading capacity) of 100 to 200 kg. The Israeli-US Hunter and UK Watchkeeper are two main examples of midsize fixed-wing. Other medium-sized UAVs used previously are, RQ-2 Pioneer, US Boeing Eagle Eye, Base Systems SkyEye RE-4E, and RQ-5A Hunter (Figures 3.18–3.21). The Hunter is 6.9 m long and has 10.2 m of wings. At launch, its total weight is around 885 kg. The American Aerospace RS-20 is another such crossover UAV that combines the features of small and

Figure 3.22 US general atomics predator b [7–9].

Figure 3.23 US northrop grumman global hawk [7–9].

medium UAVs. Apart from these, there are also a lot of medium-sized UAVs with rotary engines [5, 7–9].

3.3.1.4 Large UAVs

The military uses large UAVs primarily for combat missions, which fall within the large UAV class. There were many large UAVs available; one such large UAV is the US General Atomics Predator A and B (Figure 3.22), and

Figure 3.24 Multi-rotor UAV [7–9].

the other is the US Northrop Grumman Global Hawk (Figure 3.23) [4, 5, 7–9].

3.3.2 Classification of UAVs Based on Wing Type

3.3.2.1 Multi-Rotor Drones

Multi-rotor drones are correspondingly named 'rotary-wing' drones. These drones are one of the most popular types of drones for both personal and business use. Since they are small and easy to handle, they are an ideal choice for aerial photos (Figure 3.24).

They are very flexible and can let you put in all kinds of cameras that can do different things. These drones can easily hover and take off in any direction, giving them more options. But the primary issue with multi-rotor drones is generally the fact that they can fly on their own. When you add extra rotors to a drone, it makes it harder to control. All these moving parts also use more power, which means the battery will die faster [6]. Most drones with more than one rotor can fly for less than an hour. Long-lasting power is essential for specific jobs, so stocking up on spare batteries is essential [7–9].

3.3.2.2 Fixed-Wing Drones

Drones with fixed-wings (Figure 3.25) use their aerodynamic properties to propel themselves through the air and maintain flight. They are commonly used to survey broad regions and resemble radio-controlled airplanes in

Figure 3.25 Fixed-wing drones [7–9].

design or look [7]. Due to their superior aerodynamics and design, they can stay airborne for longer periods and travel at a faster rate.

One big drawback of fixed-wing drones is that they are generally more expensive than multi-rotor drones. Like planes, they need a big, clear area for launching and landing. Moreover, some huge fixed-wing drones additionally require special tools for takeoff and landing. Additionally, fixed-wing drones are designed to fly in one direction, so they cannot move around as well as multi-rotor drones [7–9].

Figure 3.26 Single-rotor helicopter drones [7–9].

3.3.2.3 Single-Rotor Helicopter Drones

Single-rotor drones are strong and resilient. They appear like real helicopters: they have a single rotor for propulsion and a tail for directional and stabilization control. These drones (Figure 3.26) take the best features of both multi-rotor in addition single-rotor drones, which makes them perfect for transporting heavier loads and achieving greater fuel efficiency in flight. Most single-rotor devices have gas engines instead of batteries, which give them a much longer flight time.

On the downside, these drones [8] are usually bigger and more complicated than other UAVs. This makes them more expensive, as well as more difficult to use, and because their blades are bigger, they can be more dangerous [7–9].

3.3.2.4 Fixed-Wing Hybrid VTOL Drones

Drones of the hybrid vertical take-off and landing (VTOL) type combine the best features of fixed-wing and rotor helicopter drone designs. This type of drone has rotors connected to its fixed wings. This lets it hover, take off, and land in a vertical position. Currently, there are only a few hybrid VTOL drones available in the market, but that might change as technology improves. Amazon's Prime Air-delivery drone (Figures 3.27 and 3.28) is an example of static-wing hybrid VTOL drones [7–9].

Figure 3.27 Fixed-wing hybrid VTOL drones [7–9].

Figure 3.28 Fixed-wing hybrid VTOL drones [7–9].

3.3.3 Classifications of Drones Based on Payload

3.3.3.1 Featherweight Drones

Nanoscale drones are often the name given to these aircraft, which may weigh as low as 11 g and are only capable of carrying payloads that range from 4 g up to 100 g.

3.3.3.2 Lightweight Drones

Lightweight drones are small drones that typically weigh between 200 and 1000 g, and their payload capacities typically range between 150 and 270 g.

3.3.3.3 Middleweight Drones

The term "medium-sized drones" refers to those whose weight is more than 1 kg and less than 600 kg. In general, you can split them up into two different categories: military drones and commercial drones. The payload capacity of military drones lies between 40 and 150 kg, and commercial drones may carry weights between 0.40 and 1.46 kg.

3.3.3.4 Heavy Lift Drones

Heavy lift drones (Figure 3.29) are big drones typically utilized by the military, with certain civilian applications like package delivery and precise

Figure 3.29 Heavy lift drone [7–9].

mapping. The heaviest of these drones weighs over 1,000 kg. Their load capacity is over and above 550 kg [7–9].

3.4 Security Requirements

Numerous security concerns have been brought to light by the widespread use of UAVs in commercial applications. Therefore, several security concerns are required to be addressed. They must prevent UAVs from being exposed, disrupted, modified, or destroyed. We mentioned below the following key security and privacy standards for a trustworthy UAV flight mission to ensure these security parameters are met [10]:

3.4.1 Confidentiality

Protecting private data and information exchange between a GCS and UAVs from unwanted access is essential since it might be a cause of the flight's information leaks, such as control directives and telemetry data. We need to consider developing robust cryptographic techniques to prevent adversaries from accessing information.

3.4.2 Integrity

It is crucial to maintain data integrity. Integrity is necessary for all flying missions to be successful. It holds adversaries from fabricating network communications. The UAV system's behavior could be changed if the integrity is compromised, which might cause mission failure. So, all communication must be secure and reliable. By using authenticated encryption mechanisms, we can ensure the integrity of the system.

3.4.3 Availability

Unmanned aerial vehicles must function without sabotage or malfunction. All required flight mission functions must be made available to accredited users. In addition, the system must protect UAVs from denial-of-service attacks that may be launched against them. To prevent such an attack, an IDS is employed in the system.

3.4.4 Authentication

Authentication is a vital part of making sure the UAV system can safely exchange data with itself and with stations. As part of the flight mission, it ensures that all UAVs involved in the mission are legitimate. By authenticating each UAV, we guarantee that only reliable ones participate in the aerospace mission. In addition, authentication secures the network from attackers and hackers who impersonate genuine devices.

3.4.5 Non-Repudiation

Operators cannot dispute or conceal their behavior within UAV networks, such as sending or receiving data. In the worst-case scenario of mission failure, we must address concerns regarding accountability. Non-repudiation ensures that the operator's operations are not disown under any circumstances. Also, UAV system needs to implement measures for non-repudiation, such as a digital signature for the message exchange.

3.4.6 Authorization

Access to data exchange in UAV systems should be restricted to legitimate and authorized individuals only. Additionally, the UAV system must specify which resources a valid user is authorized to access. Access control mechanisms are required to monitor the authorization of access to such resources.

3.4.7 Non-Disclosure

Information like video footage or images of sensitive areas, forwarded by UAV to a specific GCS, should not be compromised. Only the authorized GCS can access the information, so that information is not disclosed under any circumstances by any GCS or UAV.

3.5 Routing Protocols

Routing between the UAVs is critical owing to the swift motion of UAVs in FANETs, which results in a fast topology change in network. The routing protocol for UAV-to-UAV communication in FANET plays a crucial role in ensuring that data is reliably sent from one UAV to the next, and the fact that routing overhead is reduced to a minimum makes it an interesting area of study in the field of FANETs. However, the core challenge is still under investigation, i.e., in what way we can design a robust routing protocol. Conventional *ad-hoc* routing protocols are favored and researched for FANETs in the initial research and trials. However, most conventional *ad-hoc* routing protocols are not inherently appropriate for FANETs due to the special features of UAVs, for instance, as rapid connections and abrupt shifts in 3D space. Consequently, updated and modern *ad hoc* networking protocols have been proposed for routing in FANETs, leading to the development of a new family of *ad hoc* networking protocols. There are six broad groups into which these routing protocols might be placed [11]:

3.5.1 Static Routing Protocol

Each UAV in a static routing protocol uses a predetermined routing table that does not change as the journey progresses. Static routing methods can be used when the network's structure remains static and there are few viable options for routes. Each UAV shares data exclusively with the network or GCS. As a result, fewer communication links are needed for data transmission. If the update to the routing table fails, the mission must be delayed until the table is updated successfully. The protocols are, therefore, not resilient to faults.

3.5.1.1 *Load Carry and Deliver Routing*

In load and carry deliver (LCAD) routing protocol, the UAV saves data from a GCS or ground node and then flies to its target GCS or ground node to deliver this information, as shown in Figure 3.30. Although a maiden-source and maiden-destination situation was first investigated in LCAD, implementing numerous-source and numerous-destination situations is also possible if required. With less hops, this routing system is feasible for requests seeking huge data transfers and delay endurance. The LCAD routing's main goals are to enhance throughput and security. The fundamental drawback of this protocol is that the transmission latency gets excessively high and excessive once the distance between the connected UAVs grows. Several UAVs might be deployed on the identical path, with a minimal distance between them and in fast speed, to reduce transmission latency. The LCAD network may also be split up into smaller sub-networks.

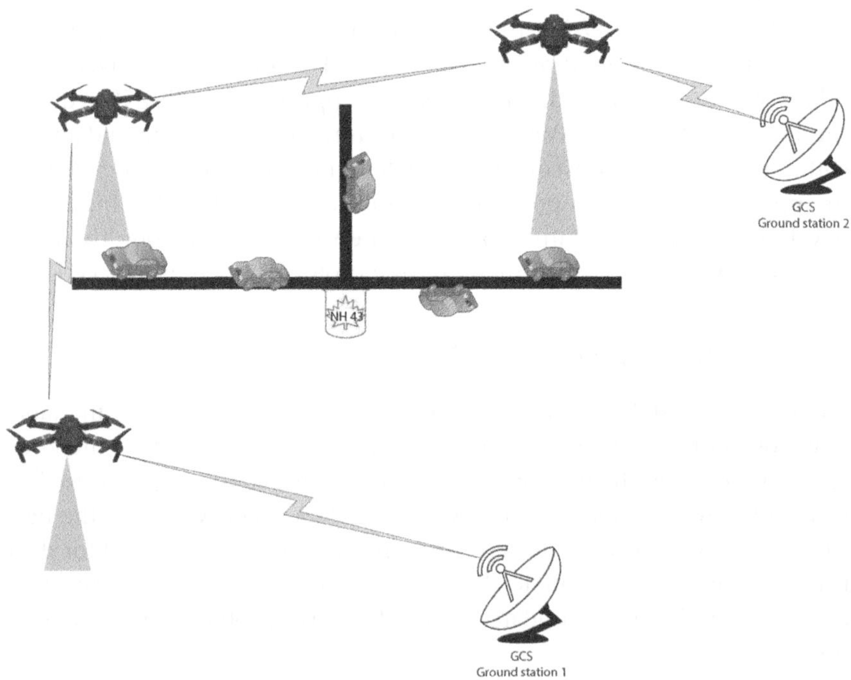

Figure 3.30 The LCAD routing protocol.

3.5.1.2 Multi-Level Hierarchical Routing Protocol

Multi-level hierarchical (MLH) protocols are another type of FANET routing techniques that work under static routing. The MLH routing protocol was designed to enumerate the problem of the network's scalability [12]. As shown in Figure 3.31, the network may be divided into many different kinds of clusters that are assigned to various operational areas. There is a cluster head (CH) at each cluster, which not only defines a full cluster but also interacts with other clusters. The MLH network permits tasks and different responsibilities to each cluster. The CH has direct communication vicinity with each UAV in a cluster. The CH has a direct or indirect connection to UAVs or satellites in the upper most layer. Better performance outcomes may be achieved by MLH protocols when many UAVs are placed in a system and all the UAVs are organized into distinct swarms with a sizable operational area. The cluster information, however, is the utmost important concern for the MLH routing. The frequent cluster data needed due to the fast mobility of UAVs was addressed by mobility prediction clustering, which used the connection expiry time mobility model and dictionary trie structure forecasting algorithm to predict the network topology information. The CH in this model is determined by selecting the UAV with the maximum weight between its neighbors' UAVs. The clusters and CHs can be more stable due to the CH selection criteria. Nadeem *et al.* [10] presented a clustering approach for UAV networks. It first designates

Figure 3.31 A multi-level hierarchical (MLH) routing protocol.

the ground-based clusters, and throughout the operation, it is updated. By using ground clustering, the clustering strategy is decided upon, and the CHs are then chosen using geographic data. Additionally, the cluster structure has been calibrated in response to mission changes just after the UAVs are deployed. This routing strategy can make dynamic networking much more stable and artistically pleasing.

3.5.1.3 *Data-Centric Routing*

As demonstrated in Figure 3.32, FANETs may implement data-centric routing (DCR) algorithms that request and collect data based on data quality instead of sender or receiver IDs. It appears that multicasting can be preferable to unicasting due to the wireless data transmission mechanism of UAVs. When many UAVs contribute to a data request, administrators can select the appropriate routing method and allow on-demand algorithms to handle data distribution. In FANETs, DCR may be employed to deliver several requests on similar multi-UAV machines to collect mission-specific data. The publish-subscribe approach often applies to such an environment [13]. It sets up an automated link between those who generate data, known as publishers, and those who consume it, known as subscribers. The producer node decides what data needs to be shared with the world and sets the ball in motion to make that happen. When the UAV in the network reaches the published data, it looks for subscription messages

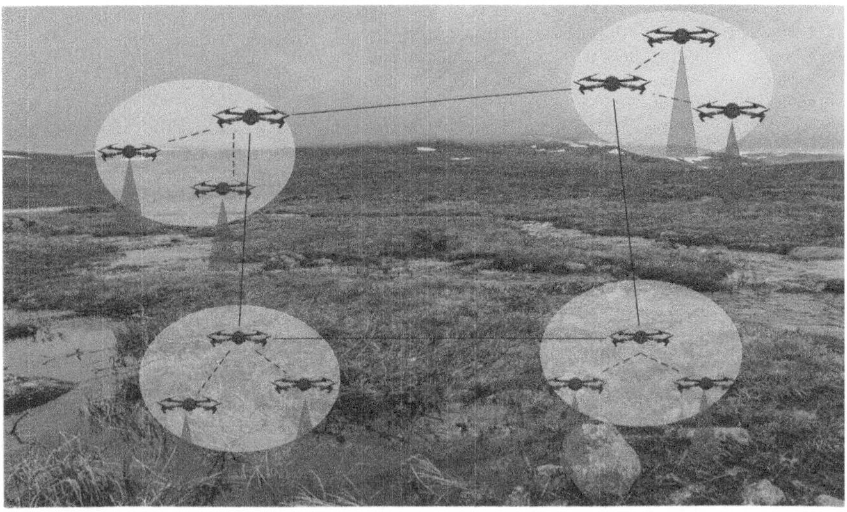

Figure 3.32 Data-centric routing.

on that data and sends them along. This routing algorithm's key strength is that it only notifies subscribers of content that has been previously registered. Algorithms for DCR are separated into three domains:

a. **Space-decoupling:** The communicating UAVs in the network can be anywhere, and all the UAVs cannot recognize the identities and geographical location of each UAVs.
b. **Time-decoupling:** The communicating UAVs in the network are not required to be constantly available, and information can be delivered to the requesting UAV either instantaneously or later.
c. **Flow-decoupling:** The transmission of information may be safely achieved through an asynchronous communication mechanism.

This algorithm works best when there are fewer UAVs in the network, and minimal assistance is required to manage the clusters.

3.5.2 Proactive Routing Protocol

In proactive routing protocols, all nodes regularly update multiple routing tables to update the nodes about the routing of the network. Proactive protocols have the upper hand in availability of pre-built routes due to their proactive nature. However, managing real time information incurs extra overhead costs, and consequently, the network's efficiency may be impacted as beacon messages are transmitted frequently even in the absence of data flow. Proactive protocols are thus not specifically suitable for dynamic, swift, and expansive networks. Further, many routing protocols exhibit a slow response to disappointment in situations when the routing topology switches, or a connection breaks. Some of the proactive protocols are explained here:

3.5.2.1 Destination Sequenced Distance Vector (DSDV) Routing Protocol

This algorithm drives from the Bellman–Ford algorithm, by some nominal changes that make it fit for the UAV network. Each UAV in the destination sequenced distance vector (DSDV) must share its data with other UAVs linked to the network. To prevent loops in routing, the routing table is periodically modernized with a sequence number for the entire network [14].

To prevent routing loops, the entire network's routing database is continuously updated with sequence numbers. Preference is given to the latest route adopted, identified by the highest sequence number, over routes with lower sequence numbers. The DSDV's simplicity and its use of sequence numbers ensure loop-free data transfer, which are its two main advantages [14]. However, a significant drawback of proactive routing is the regular updating of the current routing table, which consumes network bandwidth even if the network is idle.

3.5.2.2 Optimized Link State Routing

Optimized link state routing (OSLR) is a proactive routing protocol, so it updates its routing table regularly with new routes. As a result, the protocol instantly identifies the path to all potential destinations whenever a path is requested, then forwards that path without any initial latency. The OLSR employs a unique packet that consists of many messages to establish a communication path between UAVs running the same protocol. Three different types of messages, each serving a particular purpose, can be carried by OLSR packets. Three messages are sent periodically: the HELLO message, which is used for neighbor connectivity, link sensing, and multipoint relay (MPR) signaling; the topology control message, which advertises the periodic updating of link-state information; and the multiple interface declaration message, which permits the declaration of multiple interfaces on a node. As a result, huge overheads are produced because of this recurrent flooding feature.

Message overhead and latency can be reduced and improved with the OLSR's MPRs mechanism. This is achievable because this protocol forwards the messages only during the flooding phase. For the MPR UAVs to protect their two-hop neighbors, the sending UAV will designate a group of MPR UAVs. The MPR selector of a certain node is the UAV that chooses another UAV to be a member of the MPR UAV. Figure 3.33 depicts the MPR that the source and destination UAVs chose. The total number of MPRs that have a substantial impact on the delay is the most important design factor for OLSR. It appears that when the number of MPRs decreases, the overhead will, as well. In this way, a fresh approach to MPR membership reduction is proposed. A schematic for the proposed dynamic optimized link state routing (DOLSR) protocol is shown in Figure 3.34. The sender UAV calculates the distance between itself and the receiver UAV for each data packet it sends. If the distance exceeds the largest distance that could be reached *via* directional antenna (Dmax/2), or if an omni-directional

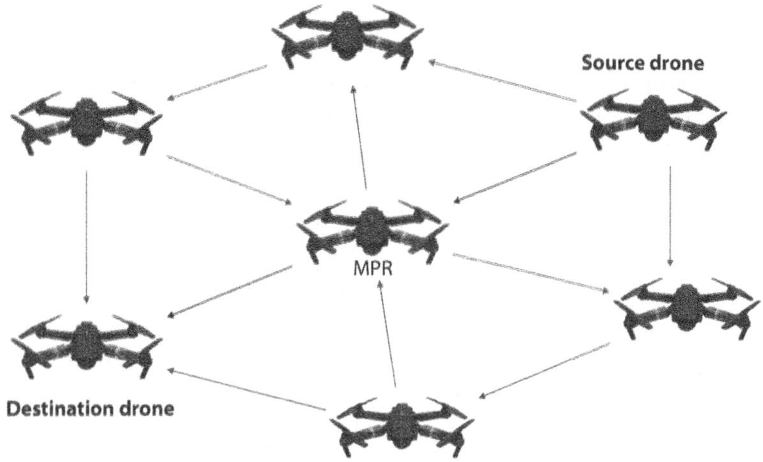

Figure 3.33 The OLSR.

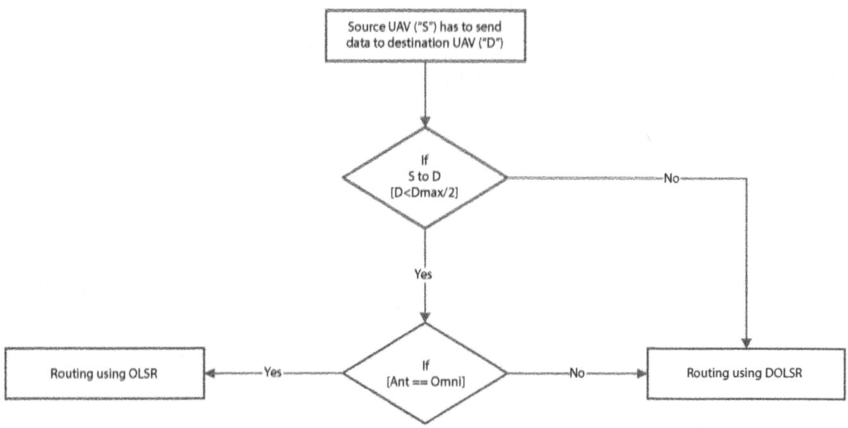

Figure 3.34 The DOLSR routing.

antenna is unable to reach the final destination, then the UAV proceeds to employ the DOLSR algorithm. If not, OLSR will often be utilized.

3.5.3 Reactive Routing Protocol

Reactive routing protocol, commonly referred to as the on-demand routing protocol, establishes and stores a route when a request is raised. A routing table is exclusively updated when a new data has to transmit. So, reactive protocols only maintain the routes which are currently in usage, which solves the proactive routing protocol overhead issue. Two different message types are generated in reactive routing model: the Route_Request message and the Route_Reply message. To find the way, the source UAV transmits the Route_Request message to all nearby UAVs using a flooding technique. Each UAV then follows the same procedure until it arrives at its destination UAV. While implementing the unicast communication method, the Route_Reply message is started by the Route_Request recipient UAV and sent to the Route_Request generator UAV. There is no requirement for the network's tables to be refreshed under this routing strategy. Since there are no scheduled updates, reactive routing techniques consume less bandwidth. The primary drawback of reactive routing is the route discovery time, which results in excessive latency in the network throughout the process of determining the best route [15].

3.5.3.1 Dynamic Source Routing Protocol

Brown *et al.* created the dynamic source routing (DSR) protocol test bed for FANETs [16]. It enables a network to operate autonomously in the absence of external infrastructure, and to configure and organize itself. The DSR is primarily used for multi-hop wireless sensor networks. Later on, it was adopted due to its reactive nature. If data must be sent, then the source in the DSR looks for a route to reach the destination in a scenario. The FANET is characterized by high UAV mobility and unstable network topology. Khare *et al.* claimed that the DSR is more efficient than proactive approaches. The extreme mobility of UAVs makes it unsuitable to update a routing table using proactive methods. Prior to the delivery of each packet, repetitive path searching using a reactive approach can be exhausting.

3.5.3.2 Ad Hoc On-Demand Distance Vector Routing Protocol

The *ad hoc* on-demand distance vector (AODV) is an enhanced version of the DSR and DSDV protocols. The AODV inherits hop-to-hop routing from DSR and periodic updates from the DSDV routing protocol. Due to its reactive nature, the AODV discovers a route only when a request is made and does not store routes to destinations that are not actively communicating [17]. The AODV protocol includes three stages: route discovery, packet transmission, and route maintenance. When the source UAV wants to transfer a packet, it first initiates a route discovery process to locate the recipient. After identifying the recipient's position, the packet is forwarded along a selected route, ensuring it does not encounter any loops. To fix connection failures, there is a phase of route maintenance.

This routing technique employs a numerical sequence number to locate the most recent, most efficient path to the desired location. To enhance the freshness of the route, a time limit is employed. The UAVs in between the two would then likewise refresh their respective trajectories. However, because of the high level of dynamically in the FANET network, network congestion and overhead are problems with AODV.

3.5.3.3 *Time-Slotted On-Demand Routing Protocol*

The AODV is essentially a time-slotted on-demand (TSOD) routing protocol [18]. Unlike the TSOD routing protocol, which allocates specific time slots for individual UAVs to transmit data packets, the AODV employs random access mode for sending control packets. This routing approach improves the efficiency of the usable bandwidth while simultaneously reducing packet collisions and improving packet delivery rates.

3.5.4 Hybrid Routing Protocols

Hybrid routing protocols combine proactive and reactive routing protocols to improve upon their respective strengths and address their weaknesses. Proactive routing methods have a high overhead of control messages, whereas reactive ones take longer to find a route. The use of hybrid routing protocols can help make up for these drawbacks. Hybrid protocols are ideally suitable for large networks; they utilize zone concepts, where proactive routing protocols are used for intra-zone routing and reactive routing protocols are used for inner-zone routing.

3.5.4.1 *Zone Routing Protocol*

Zone routing protocol (ZRP) primarily relies on the principle of "zones." This protocol suits extensive network spans and diverse mobility patterns. The UAVs in this approach have a unique zone, with overlapping zones among nearby UAVs. The radius of size "R" determines the zone's size. The distance "R" represents the distance in UAVs from the zone's edge. By adjusting the transmission power of UAVs, it becomes feasible to manage numerous UAVs within a specified area. Intra-zone routing denotes the process of routing exclusively within a zone. The "proactive routing" strategy is used by intra-zone routing to preserve the routes [19]. The source UAV can begin data communication right away if the destination and source UAVs are in the same zone. Sending data packets beyond the zone is now the responsibility of inter-zone routing, which uses reactive routing to select and maintain the best paths. Border-casting is utilized to reduce route discovery time. Only a zone's edge/border UAVs can create replies. The edge UAVs then repeat, choosing either an intra- or inter-zone route based on what is necessary.

3.5.4.2 Temporarily Ordered Routing Algorithm

Temporarily ordered routing algorithm (TORA) is a versatile on-demand routing technique that performs well in multi-hop networks [20]. Each UAV solely updates route information for nearby UAVs in this routing method. A significant benefit of employing TORA is that it limits the spread of control messages in an environment with high mobility to reduce speedy responses to topological shifts. In a single run of the distributed algorithm, it removes useless routes and explores fresh routes. This protocol uses reactive routing protocols, although it also uses proactive methods on occasion. A directed acyclic graph (DAG) is built and maintained between the source and destination UAVs. The UAVs in DAG travel a variety of routes. This is helpful in expediting the computation of fresh routes if links become disconnected and for improving adaptability. The TORA does not rely on the shortest path; instead, it often chooses longer paths to minimize the network overhead. There is a constraint value called "height" for each UAV in the DAG, which defines that no two UAVs are at the same height. A top-down technique is used in the data transfer from higher UAVs to lower UAVs. Due to the lack of data streams toward higher-height UAVs, it provides loop-free routing. This height variable is sent back to the inquiring UAV during the route discovery process. In this method, all intermediate UAVs keep their routing tables by receiving routes and height data.

3.5.5 Geographic-Based Routing Protocols

Position-verification routing protocols assume the geographic location of UAVs to facilitate more efficient routing. This class of protocols assume that the source UAV possess knowledge of the physical locations of the other communicating UAVs, enabling direct communication with destination UAVs without the need for initial path discovery. Typically, each UAV uses a GPS or another sort of positioning facility to identify its own location. The main sources of inspiration for this routing system are two things [21]: first, utilizing position tracking by the data packet sender to pinpoint the receiver's exact location; and second, employing a forwarding mechanism to relay data packets to the intended UAV target.

3.5.5.1 Greedy Perimeter Stateless Routing

A greedy routing protocol is a position-based protocol that outperforms proactive and reactive routing protocols in terms of performance. A simulation framework was created by Shirani *et al.* [54] to analyze position-based routing protocols in FANETs. Studies have shown that greedy geographic-driven forwarding routing protocols are suitable for densely distributed UAV networks. However, in sparsely populated installations, ensuring network reliability may present a significant challenge. Applications that aim for 100% dependability need to include many other strategies, such as face routing. The majority of the conventional MANET routing algorithms, which have initially been evaluated for FANET deployments, are now inappropriate for FANETs because of specific UAV obstructions including fast changes in connection quality and extremely high-flying node mobility.

3.5.5.2 Mobility-Oriented Geographical Routing

The concept of geographic position mobility-oriented routing (GPMOR) was introduced for FANETs [55]. Unlike traditional position-based systems that rely solely on UAV locations, GPMOR incorporates the "Gaussian–Markov mobility model" to predict UAV trajectories. This predictive capability enables GPMOR to determine the next hop for routing data packets efficiently. The effectiveness of this routing system is evaluated based on latency and packet-delivery ratio, aiming to minimize data loss during transmission.

3.5.6 Hierarchical Routing Protocols

Hierarchical routing protocols allow a UAV to switch between proactive and reactive routing depending on its position in the network's hierarchy. This routing approach relies heavily on predefined proactive routes, which are then utilized to handle requests from other UAVs using reactive protocols at lower tiers. However, the protocol's complexity and its addressing method, which responds to traffic requests but complicates the interconnection of components, remain significant drawbacks.

3.5.7 Mobility Prediction Clustering Algorithm

Mobility prediction clustering algorithm [22] is based on the characteristics of UAVs that have been developed for use in UAV networking. To address the challenges relating to the high mobility of UAVs, this protocol relies on trie-structure forecasting algorithm's dictionary and connection termination time. This technique enhances network performance by decreasing clustering instability.

3.5.8 Clustering Algorithm

The clustering mechanism was designed as a solution to the challenge of managing UAVs, which are hidden from view. In the event of a multi-UAV system, this algorithm first constructs the clusters on the surface of the earth and then adjusts them in space, with the goal of enhancing the stability and adaptability of the near-space clusters.

3.6 Security Issues and Countermeasures in FANET

Security in FANET is a primary concern as it deals with real-time data forwarded to aerospace to take real-time action on the received data. Therefore, it is important to secure the network from adversaries and attacks. In this section, we list some security issues in FANET. Here, we categorized the security issues into the following domains [23]: sensor-level security issues, hardware-level security issues, and software-level security issues.

3.6.1 Sensor Level Security Issues

The UAVs are dependent on several sensors to collect data of their surroundings. Since the collected data is sensitive, it is necessary to protect it from attackers. In hostile conditions, compromised UAV sensors could cause UAV systems to fail. In the subsequent sections, we examine various sensor-level vulnerabilities, associated risks, and suggested countermeasures.

3.6.1.1 Vulnerabilities and Treats

The UAVs are incredibly sensor-driven gadgets. They have a range of sensors, including cameras, motion sensors, GPS, LiDAR, and accelerometers. As a result, they completely depend on the data collected by the sensor to function properly. However, because these sensors deal with private data, a malicious user could use them to sabotage the flight mission. For instance, civil GPS transmissions lack both authentication and encryption. As a result, an attacker can take advantage of this loophole by fooling the operator with a fake GPS signal. The UAV system could fail if an attacker used real-time data from onboard sensors. This vulnerability exists because the air traffic controller does not verify the accuracy of sensor data. Additionally, malicious software may be used to introduce sensor-level vulnerabilities into the system. This class of vulnerabilities introduced a new attack domain, enabling the adversary to totally control commercial aircraft.

3.6.1.2 *Sensor-Based Attacks*

- **Global positioning system data jamming:** During flight missions, the on-board GPS sensor acquires the location of aircraft and drones from satellites and transmits it to the GCS. The GPS jamming attacks occur when an adversary disrupts the GPS signals, leaving the UAV disoriented [24]. As a result, the UAV loses its control and makes the drones susceptible to hijacking.
- **Substituted sensor data:** External sensors like infrared sensors and electro optical sensors can get jeopardized by inserting a fraudulent sensor's information in a flight controller (FC). This attack causes a malfunction in the UAV system. The on-board aircraft controller system may be accessed by an attacker, or the attacker can change the sensor data through system calls, to inject fake sensor's information into UAVs. On the contrary, the attacker may send fake signals directly into FANET environment devices and make the UAVs vulnerable. Global positioning system spoofing serves as a prominent example of such attacks, involving the insertion of misleading sensor data. Since GPS signal broadcasts are often unencrypted and unauthenticated, an attacker may launch a spoofing attack by generating a signal that seems to come from the GPS, thereby causing the GPS receiver of UAVs to be compromised. Consequently, the UAV is now under the attacker's command.
- **Sensory channel attack:** UAVs rely on several sensors, the channels of which might be exploited for malicious purposes. The deliberate noise may cause malfunctions in UAVs fitted with gyroscopes made by micro-electro-mechanical systems (MEMS) [25]. Findings [26] demonstrate that MEMS gyroscopes work on audible resonant frequencies. The camera sensor used to examine UAVs may get effected because of the environment's effects [27].

3.6.1.3 Defense Mechanisms Against Sensor-Based Attacks

To prevent GPS jamming strikes, Kim and Lee [28] came up with an approach for the flight controller to navigate without GPS signals. The authors [29] proposed some different approaches based on machine learning and intrusion detection systems. Authors gathered training datasets from on-board UAV components, including flight records and sensor readings, and has a F1 statistic of 99.73% for fraudulent sensor readings across many UAV platforms [30]. However, real-world applications are difficult to implement and execute because of the UAVs' limited energy and computational power. De Rango *et al.* [31] suggested using an extra navigation sensor to help with the tracking of UAVs when GPS signals are not available. The authors' solutions for UAV autonomy in a GPS-loss scenario was to pair an inertial measurement unit (IMU) sensor with a monocular camera visual sensor.

We may cross-verify the sensor data by collecting readings from a different set of sensors to stop the injection of false flight controller sensor data. Analyzing the UAV's physical attributes using the control invariant approach is another method for detecting external sensor threats [32]. The control-invariant method ensures that the actual state of the UAV matches the reading predicted by the control model. The proposed method has been tested and shown to identify sensor attacks in under a hundred milliseconds in evaluations involving many quadrotors. In a similar vein, the author proposed an architecture for protecting sensor data in the face of physical invariants. As distinct characteristics, the UAVs' physical invariants may be modeled to anticipate sensor data. This functionality is represented by a set of non-linear differential equations that model the UAV's velocity, orientation, location, and angular velocity. The results reveal that sensor-based attacks can potentially be detected in under 100ms with a false alarm rate < 2% by using well-known physical invariants to learn their characteristics.

Overcoming an adversary's GPS spoofing attack might be as simple as monitoring for power fluctuations in the signal to warn of a forthcoming attack. Souza *et al.* [33] suggested an integrated data attestation system for multi-UAV situations that certifies the accuracy of shared data, including GPS coordinates, and can identify GPS spoofing attacks. This framework is built on the concept of control flow graph (CFG). According to the framework's performance over the PixHawk UAV's GPS sensor, the generation times of the attestation report are 835 ms and the verification time is 849 ms, respectively, for a CFG size of 2922.

Garcia–Santiago *et al.* [34] suggested an information fusion mechanism, which is the combination of an IMU sensor and a camera, to identify GPS spoofing attacks on the devices in the network. The experimental result shows the success of this method only when the x-axis is greater than 2,046 ms and the y-axis is greater than 23,311 ms during the flight. The other defense mechanism is to authenticate GPS signals using a cryptographic approach to eliminate GPS spoofing attacks. This approach is costlier as it requires additional support and hardware [34]. In our literature review, we also observed that certain anti-GPS spoofing techniques were appropriate for integration into the flight controller, which is capable enough to efficiently handle hacking. One such technique is the GPS spoofing detection approach [35], which is the combination of GPS and IMU to identify GPS spoofing. For instance, Grasso *et al.* [36] suggested using GPS and IMU data to identify GPS faking. This method uses genetic algorithm in conjunction with a two-step ML model. To minimize the cost of computing, the model first pre-trained the drone. To attain a high detection rate, the model is then trained on the drone. The simulation shows 96% and 100% accuracy of the model in hijacked and non-hijacked scenarios.

In the literature review, a number of mechanisms have been proposed to counter multiple sensory-channel attacks. Physical isolation, which may mitigate noise, protects acoustic sensory channels [37]. To determine visual motion, optical flow sensors are completely dependent on optical flow algorithms. Creating strong optical flow algorithms is a way to protect optical flow devices from being manipulated.

3.6.2 Hardware Level Issues

Unmanned aerial vehicles are perceived as potential tools for adversaries to execute direct physical attacks within airspace. The UAV system typically consists of a ground base station and a flight controller. Security vulnerabilities present in both hardware components have the potential to facilitate physical or cyber-attacks. We delineate specific vulnerabilities in these areas that attackers could exploit to compromise UAV hardware, followed by an exploration of present defense strategies aimed at mitigating such hardware-based attacks.

3.6.2.1 Vulnerabilities and Threats

- **Hardware trojans:** This threat affects electronic devices, for instance by altering logic gate else altering the device circuit. The FC is a prime target of hardware trojans, which opens the UAV system up to a wide variety of threats. A malicious and unreliable third party inserted hardware trojans in the FC's semiconductor source chain. The attacker practices such vulnerabilities to breach the security and functionality of the FC's IC, i.e., by leaking the FC's cryptographic keys and decelerating propeller spin. One such trojan was discovered in the ActelProASIC chip used in the Boeing 787 plane [38]. The backdoor posed a significant risk to the flight mission's safety, allowing unauthorized control of the UAV by the attacker.

- **Physical UAV collision:** During complex flying missions involving multiple UAVs, physical collisions leading to UAV collapse may occur, highlighting the need for enhanced coordination and cooperation strategies. The UAVs mainly depend on a collision avoidance scheme to evade accidents in airspace. The public UAV systems, however, lack integrated security features and cannot address the problem of hostile agents' collision avoidance.
- **Hardware failure:** Hardware malfunctions like battery degradation and motor issues can jeopardize UAV missions, leading to emergency landings in unexpected locations. If UAVs transmit unencrypted data in these scenarios, then adversaries could exploit this vulnerability to compromise classified mission information, breaching mission confidentiality and integrity.
- **Flying skill issues:** Operational challenges arise with non- and semi-autonomous UAVs commanded remotely by humans, particularly with large and complex UAVs [39]. The effective management of the UAV's velocity, altitude, and path by FCs requires operators with sufficient technical proficiency. Inadequate expertise increases the risk of drone crashes and mission failure, potentially resulting in UAV theft.

3.6.2.2 Hardware-Based Attacks

- **Hijacking:** UAVs are the soft targets for hijacking because of their propensity to be observable at low altitude. An enemy can take control of a flying drone in one of two ways: first is directly, second is remotely, using malicious tools. Anti-drone guns, typically wielded by law enforcement to monitor unauthorized UAVs in restricted areas, can also be used by attackers to disable drones and commandeer them for nefarious purposes.

- **Supply chain attack:** The rapid development of the drone industry has created new opportunities for adversaries to jeopardize UAVs through supply chain attacks. These attacks focus on exploiting vulnerabilities in critical components like airframes, actuators, and propellers during the organization's supply chain process. This compromises the integrity of the final product delivered to consumers. Researchers explored effective strategies for targeting UAV supply chains using additive manufacturing [40]. They demonstrated how attackers could remotely manipulate propeller design documents to sabotage specific UAVs, diminishing their endurance and operational effectiveness. This research underscores the difficulty in detecting and preventing sabotage within additive manufacturing processes.
- **Battery attack:** Rechargeable lithium-ion batteries are commonly used to power UAVs, supported by a battery management system that ensures stable power distribution to various UAV components. However, adversaries may exploit vulnerabilities through battery depletion attacks, leading to UAV malfunctions and compromising battery availability, integrity, and confidentiality. Attackers can disrupt the UAV's power supply by physically tampering with or replacing reliable batteries with defective ones. Additionally, deep discharge attacks on batteries can compromise other UAV components, such as manipulating sensors or installing malicious software [41]. Another method to target UAV battery security involves modifying genuine battery information exchanged between UAVs and GCS, potentially exposing sensitive data like state-of-charge and undermining UAV battery privacy.
- **Radio frequency (RF) module attack:** RF modules enable devices to exchange radio messages. In UAV operations, users typically issue control commands to drones *via* a GCS or a remote controller. Disruption of the UAV-to-GCS connection or jamming of control signals can result in a lost-link scenario. Leonov and Litvinov [42] demonstrated a replay attack on the XBee 868LP protocol, used in UAV-to-GCS communication with low-power RF modules. This attack

involves manipulating transmissions between the UAV and GCS using a third XBee chip. By exploiting the XBee channel address through existing chip features, the attacker compromises the confidentiality of the primary XBee communication channel.

3.6.2.3 Defense Mechanisms Against Hardware-Based Attack

Physical protection strategies should be taken into consideration and improved to counter the risks caused by UAVs' physical weaknesses. Building ML-based IDS to identify such hardware attacks is one of the proposed hardware trojan mitigation strategies to ensure a trojan-free drone. Intrusion detection systems solutions work. First is developing a model from the average data produced by the pulse width modulation (PWM) signals, which is then used to detect the existence of tampered data or signals. These signals are frequently implemented inside the integrated circuits of UAVs. Second, training the machine with malicious data. By introducing hardware trojans or hacking the firmware, these data are produced, which result in an effect on PWM signals. Another method to prevent attacks is to carry out fine-grained circuit examinations to identify hardware trojans [43]. In this situation, a probability of higher than 0.99 may be attained for trojan identification over a wide range of configurations.

Malicious attackers will be substantially less likely to take control of and hijack the flying UAV if the UAVs and GCS are kept malware-free and protected from unauthorized access *via* authenticated encryption. Additionally, altering the flight routes could make it more difficult for the adversary to figure out the flying pattern, which would make it more difficult for the attacker to physically steal the UAVs. A UAV hijacking detection method [44], which is based on statistical study of typical flight patterns was released. This detection strategy is effective against over 20 hijacking cases in more than 50 simulated baseline flights, as demonstrated by the modeling of various hijacking scenarios. However, when simulation variables like control instability vary, their algorithm fails, which encourages additional testing and the enhancement of the simulation data's quality.

Avoiding the use of compromised UAV components is one way to reduce the risk of supply chain attacks [45]. This can be achieved by careful supply chain security management throughout production. Additionally, tamper-proofing methods, such as anti-tamper software and tamper-proof microprocessors, will prevent illegal physical or logical manipulations that can compromise the reliability of the UAV's essential components.

The battery management system's safety circuits guarantee physical battery security for UAVs and are one of the existing solutions to counteract battery depletion attacks [46]. Additionally, a preflight inspection of batteries would be a fair procedure to ensure a safe flight operation.

For instance, the Parrot AR drone was used for the experiment's results. The result signifies that the true lifespan of defective batteries is significantly less than that of regular batteries under battery replacement attack. Another method, which involves the real-time continuous tracking of the battery draining procedure, might potentially be used to find depletion attacks during the flying operation. In contrast, if UAV-2-GCS data transfer is not verified, the attacker may forge the message and mislead the operator about the battery level. As a result, we must use cryptographic techniques to protect UAV-2-GCS data transfers. Additionally, by employing attributes collected from simulated battery starvation attacks, we may use machine learning approaches to automatically predict battery depletion attacks in UAVs [46].

The chip's onboard encryption can be used by the manufacturer to reduce RF module threats. Due to the chip's higher latency and reduced capacity, this approach is still constrained. A solution [47] suggests exporting the encryption to a second distinct chip in this situation. Since the commands are transferred directly to the second chip, this technique ensures the privacy of the data transmitted over the radio network but does not stop the attacker from executing distant commands. Therefore, by assigning arbitrary values to destination addresses, the adversary can launch a DoS attack. The encryption of radio control channels is an alternative method to consider. Based on the Galois Embedded Crypto package [47], developers [48] created an encrypted RC link that is suitable for devices with limited resources. The recommended approach ensures safe communication between the RC transmitter and the UAV. By adopting an auditory detecting mechanism based on microphones, Lakhwani *et al.* [49] developed a prototype for protecting UAVs from physical attacks involving moving items. The solution considers a short-term Fourier transform algorithm that gives UAVs the ability to dodge objects and identify them in 100 milliseconds.

3.6.3 Software Level Issues

After discussing sensor and hardware-level issues, we learn about vulnerabilities, software-level issues, and threats in this session. Later, we will illustrate some defense mechanisms.

3.6.3.1 *Vulnerabilities and Threats*

- **Malicious software**

Malicious software presents significant risks to both FCs and GCSs in UAV operations. The UAV malware can result in the compromise of confidential information and control over deployed UAV systems. If attackers infiltrate the GCS, they could deactivate the system, leading to a DoS and disrupting missions. Embedded malware poses serious threats to the security and privacy of UAVs. For example, Maldrone [46] is a program that compromises FCs, granting attackers control over UAV systems. This malware intercepts communications between the drone's FC and sensors, enabling attackers to direct the compromised drone's landing. Another threat is the SkyJack virus [50], which uses Wi-Fi de-authentication attacks to remotely seize control of authorized drones and compromise entire systems. Snoopy spyware [51] can infect drones and steal personal information by tricking users into connecting to fake Wi-Fi networks through impersonation tactics. Recently, a new type of malware known as ransomware [52] has emerged, although UAVs have not yet been reported as targets. Given the widespread use of UAVs, it is essential to recognize the potential for future ransomware attacks against UAV systems.

- **Zero-day vulnerabilities**

Both the flying UAV and the GCS software may harbor unidentified vulnerabilities, of which manufacturers are unaware, posing serious risks to users. Adversaries can exploit these zero-day vulnerabilities until UAV developers issue updates. Operators must promptly update their UAV systems each time a patch is released to mitigate these risks.

3.6.3.2 *Software-Level Attacks*

- **Operating system attacks**

The system software of the FC is vulnerable to attacks in both military and civilian missions, potentially resulting in the destruction of UAV systems and loss of critical data. An example of such vulnerability was demonstrated in the targeting of Amazon Prime Air's parcel delivery UAVs [53]. A successful attack on the delivery system could lead to the loss of the shipment and the subsequent crash of the drone during its mission. Attacks on UAV operating systems (OS) typically involve remotely injecting malicious software, such as Maldrone [54] into UAVs to gain control and hijack the drone. This allows adversaries to exploit vulnerabilities and access unencrypted data stored in the FC by obtaining the FC's cryptographic data.

- **Tampering captured video**

System calls within the OS enable recording from the onboard camera, crucial for safe navigation and collision prevention during UAV missions [55]. However, skilled hackers with system access can intercept these calls to assume control of UAVs. Combining tampering with GPS spoofing allows attackers to commandeer UAVs during flight, aiming to compromise navigation safety and provoke collisions.

- **System ID spoofing**

FAA guidelines [56] say that UAVs should tell third parties, like government agencies and law enforcement, their system ID, and position when requested. However, because the majority of current UAVs lack encryption protections, attackers can pretend to be legitimate third parties and carry out spoofing attacks to break the communication link and get a UAV's system identity.

3.6.3.3 *Defense Mechanism Against Software-Based Attack*

Regular OS updates can protect UAVs and their payloads. Additionally, the GCS's firewall systems can prevent the transmission of harmful traffic to the flying drones. Furthermore, software-based solutions (i.e. IDS and antivirus) monitor network traffic and protect UAVs from malicious activity. For UAV networks, Cumino *et al.* [57] proposed an IDS. Whereas the simulation results show an impressive detection efficiency of over 93% and a low proportion of false positives of under 3%, updating the total count of UAVs would greatly raise the percentage of false negatives and consumption of energy, which can have an immediate impact on the network's scalability. Also, activating the authorization methods for UAV resources can aid in preventing the execution of malicious code. Employing software-based attestation techniques is a potential defense against assaults that exploit software. They make sure the software operating on the flight stack is trustworthy [58]. The software stack has a good level of legitimacy because of remote attestation technologies, which are not so expensive. The Secure Asynchronous Remote Attestation (SARA) protocol conducts attestation over a significant number of IoT devices [59]. It has a small storage consumption of 3.03 KB, an execution time of 19 sec for 250 requests, and a small energy usage of 0.196mJ, as realistic simulations showed [60].

The adversary uses malicious malware and zero-day vulnerabilities at the software level to attack the flight stack. Additionally, the attacker might tamper with the recorded footage to deceive the operator. Antivirus software and IDS can reduce or prevent these types of attacks from happening.

To confirm the authenticity of the code executing on the OS, the operator should also install software-based attestation techniques and keep the OS updated. It is important to note that the protection measures against software-based threats fall short of completely shielding the flight stack from harmful actions. For publicly publicized zero-day vulnerabilities, the patching procedure might take a long time. Consequently, the UAVs become exposed to attackers. Additionally, deploying IDS or firewall systems on the GCS might raise the cost of computation and impair latency.

3.7 Conclusion

The objective of the proposed chapter is to present a comprehensive description of current FANET security risks, countermeasures, and pertinent methodologies. We classify the architecture, type, and connection mechanism of UAVs with GCS. Then we classify UAVs based on size, type, and payload capacity. Multiple routing protocols, are then classified, which resolve certain problems and enhance FANET usage. Most of the classification and mechanisms of UAVs are illustrated using diagrams and simple illustrations. Later, we classify the security requirements for FANET, which were adapted by academia, business, and standards bodies. Further, three major security issues: sensor level security issues, hardware level issues, and software level issues, are observed and illustrated in this chapter. They also suggest countermeasures against the potential attacks or issues over FANET. Flying *ad hoc* network security solutions were previously covered in multiple surveys; however, we also covered alternative solutions. Both conventional solutions and recent FANET technologies and protocols are included in this survey. The purpose of this study was to present concise theories underlying the development of security risks, their related security impacts, and their appropriate remedies. Our target readers seek comprehensive understanding about FANET and UAV security, privacy issues and appropriate solutions. For anyone looking to learn more about mobile vehicles or begin a new initiative in this area, for instance, reading this chapter would be helpful. We also covered current regulations for UAV design, safety, communication, traffic control, and human training. To identify or protect against security risks, UAVs or GCSs can be equipped with agent-based or agent-less systems. Without the use of an effective machine learning method, many UAV types may not be suited for agent-based solutions due to hardware restrictions. In our opinion, future research might combine agent-less and agent-based systems. Many current solutions try to address difficulties with jamming, excessive latency, and poor network quality. Future research may concentrate on security measures that are more effective with less processing power, such as software-defined networks, reinforcement learning, and lightweight encryption. The FANET defense strategies scheme should consider its limitations—size, vastly mobile UAVs, major conviction on radio propagation, frequent topology changes, and constrained processing abilities—instead of applying MANET or VANET features. We think that the development of FANET and IoD defense systems is still in its infancy, and we eagerly await more attention and advancement in this field.

References

1. Bujari, A., Palazzi, C.E., Ronzani, D., FANET Application Scenarios and Mobility Models, in: *Proceedings of the 3rd Workshop on Micro Aerial Vehicle Networks, Systems, and Applications (DroNet '17)*, Association for Computing Machinery, New York, NY, USA, pp. 43–46, 2017, https://doi.org/10.1145/3086439.3086440.
2. Srivastava, A. and Prakash, J., Future FANET with application and enabling techniques: Anatomization and sustainability issues. *Comput. Sci. Rev.*, 39, 100359, 2021, ISSN 1574-0137, https://doi.org/10.1016/j.cosrev.2020.100359.
3. Chriki, A., Touati, H., Snoussi, H., Kamoun, F., FANET: Communication, mobility models and security issues. *Comput. Netw.*, 163, 106877, 2019, ISSN 1389-1286, https://doi.org/10.1016/j.comnet.2019.106877.
4. Mahmud, I. and Cho, Y.-Z., Adaptive Hello Interval in FANET Routing Protocols for Green UAVs. *IEEE Access*, 7, 63004–63015, 2019, doi: 10.1109/ACCESS.2019.2917075.
5. Bekmezci, I., Sen, I., Erkalkan, E., Flying *ad hoc* networks (FANET) test bed implementation. *2015 7th International Conference on Recent Advances in Space Technologies (RAST)*, Istanbul, Turkey, pp. 665–668, 2015, doi: 10.1109/RAST.2015.7208426.
6. Mowla, N.I., Tran, N.H., Doh, I., Chae, K., AFRL: Adaptive federated reinforcement learning for intelligent jamming defense in FANET. *J. Commun. Netw.*, 22, 3, 244–258, June 2020, doi: 10.1109/JCN.2020.000015.
7. https://www.ga-asi.com/
8. https://www.arrive.tech/dronedek-is-now-arrive/
9. https://www.directindustry.com/
10. Nadeem, A., Alghamdi, T., Yawar, A., Mehmood, A., Siddiqui, M.S., A Review and Classification of Flying *Ad-Hoc* Network (FANET) Routing Strategies. *J. Basic Appl. Sci. Res.*, 8, 3, 1–8, 2018, ISSN 2090-4304.
11. Parihar, A.S. and Chakraborty, S.K., A simple R-UAV permission-based distributed mutual exclusion in FANET. *Wirel. Netw.*, 28, 779–795, 2022, https://doi.org/10.1007/s11276-022-02889-y.

12. Parihar, A.S. and Chakraborty, S.K., Token Based k-Mutual Exclusion for Multi-UAV FANET. *Wirel. Pers. Commun.*, 126, 3693–3714, 2022, https://doi.org/10.1007/s11277-022-09886-6.
13. Khan, M.A., Qureshi, I.M., Khanzada, F., A Hybrid Communication Scheme for Efficient and Low-Cost Deployment of Future Flying *Ad-Hoc* Network (FANET). *Drones*, 3, 16, 2019, https://doi.org/10.3390/drones3010016.
14. Sun, G., Qin, D., Lan, T., Ma, L., Research on Clustering Routing Protocol Based on Improved PSO in FANET. *IEEE Sens. J.*, 21, 23, 27168–27185, 1 Dec. 1, 2021, doi: 10.1109/JSEN.2021.3117496.
15. Rajasekaran, A.S., Maria, A., Al-Turjman, F., Altrjman, C., Mostarda, L., Anonymous Mutual and Batch Authentication with Location Privacy of UAV in FANET. *Drones*, 6, 14, 2022, https://doi.org/10.3390/drones6010014.
16. Colajanni, G., Daniele, P., Galluccio, L., Grasso, C., Schembra, G., Service Chain Placement Optimization in 5G FANET-Based Network Edge. *IEEE Commun. Mag.*, 60, 11, 60–65, November 2022, doi: 10.1109/MCOM.001.2200014.
17. Liu, C. and Zhang, Z., Towards a robust FANET: Distributed node importance estimation-based connectivity maintenance for UAV swarms. *Ad Hoc Networks*, 125, 102734, 2022, ISSN 1570-8705, https://doi.org/10.1016/j.adhoc.2021.102734.
18. Li, X. and Huang, J., ABPP: An Adaptive Beacon Scheme for Geographic Routing in FANET. *2017 18th International Conference on Parallel and Distributed Computing, Applications and Technologies (PDCAT)*, Taipei, Taiwan, pp. 293–299, 2017, doi: 10.1109/PDCAT.2017.00055.
19. Faisal, S.M. and Vajpayee, A.K., Extended Zone Routing Protocol. *Int. J. Comput. Sci. Eng.*, 5, 5, 151–154, 2017.
20. Zaidi, T. and Faisal, S., An overview: Various attacks in VANET. *4th International Conference on Computing Communication and Automation (ICCCA)*, pp. 1–6, 2018.
21. Ayub, M.S., Adasme, P., Melgarejo, D.C., Rosa, R.L., Rodríguez, D.Z., Intelligent Hello Dissemination Model for FANET Routing Protocols. *IEEE Access*, 10, 46513–46525, 2022, doi: 10.1109/ACCESS.2022.3170066.
22. Khan, S. and Khan, M.Z., An Ant-Hocnet Routing Protocol Based on Optimized Fuzzy Logic for Swarm of UAVs in FANET. *Wirel. Commun. Mobile Comput.*, Hindawi, 2022, 12, 2022, https://doi.org/10.1155/2022/6783777.
23. Palan, K. and Sharma, P., FANET Communication Protocols: A Survey. *IJARCCE*, 7, 2, 219–223, Sept 2015 - March 2016.
24. Beegum, T.R., Idris, M.Y.I., Ayub, M.N.B., Shehadeh, H.A., Optimized Routing of UAVs Using Bio-Inspired Algorithm in FANET: A Systematic Review. *IEEE Access*, 11, 15588–15622, 2023, doi: 10.1109/ACCESS.2023.3244067.
25. Ahmed Sumra, I., Sellappan, P., Abdullah, A., Ali, A., Security issues and Challenges in MANET-VANET-FANET: A Survey. *EAI Endorsed Trans.*

Energy Web [Internet], 5, 17, 6, 2018, https://publications.eai.eu/index.php/ew/article/view/1011.

26. Maistrenko, V.A., Alexey, L.V., Danil, V.A., Experimental estimate of using the ant colony optimization algorithm to solve the routing problem in FANET. *2016 International Siberian Conference on Control and Communications (SIBCON)*, Moscow, Russia, pp. 1–10, 2016, doi: 10.1109/SIBCON.2016.7491805.
27. Cappello, G.M., *et al.*, Optimizing FANET Lifetime for 5G Softwarized Network Provisioning. *IEEE Trans. Netw. Serv. Manage.*, 19, 4, 4629–4649, Dec. 2022, doi: 10.1109/TNSM.2022.3193883.
28. Kim, D.-Y. and Lee, J.-W., Joint Mission Assignment and Topology Management in the Mission-Critical FANET. *IEEE Internet Things J.*, 7, 3, 2368–2385, March 2020, doi: 10.1109/JIOT.2019.2958130.
29. Fixed-wing UAV - BramorsAR - C-Astral - observation/autonomous/radio-controlled (directindustry.com).
30. Pandey, A., Shukla, P.K., Agrawal, R., An adaptive Flying *Ad-hoc* Network (FANET) for disaster response operations to improve quality of service (QoS). *Mod. Phys. Lett. B*, 34, 10, 2050010, 2020.
31. De Rango, F., Potrino, G., Tropea, M., Santamaria, A.F., Fazio, P., Scalable and ligthway bio-inspired coordination protocol for FANET in precision agriculture applications. *Comput. Electr. Eng.*, 74, 305–318, 2019, ISSN 0045-7906, https://doi.org/10.1016/j.compeleceng.2019.01.018.
32. Bekmezci, I., Sahingoz, O.K., Temel, S., Flying *Ad-Hoc* Networks (FANETs): A survey. *Ad Hoc Netw.*, 11, 3, 1254–1270, 2013, ISSN 1570-8705, https://doi.org/10.1016/j.adhoc.2012.12.004.
33. Jorge, S., José, J., Tássio, C., Jasmine, A., Renato, F., A Proposal for Routing Protocol for FANET: A Fuzzy System Approach with QoE/QoS Guarantee. *Wirel. Commun. Mobile Comput.*, Hindawi, 2019, 10, 2019. ISSN 1530-8669, https://doi.org/10.1155/2019/8709249.
34. Garcia-Santiago, A., Castaneda-Camacho, J., Guerrero-Castellanos, J.F., Mino-Aguilar, G., Evaluation of AODV and DSDV routing protocols for a FANET: Further results towards robotic vehicle networks. *2018 IEEE 9th Latin American Symposium on Circuits & Systems (LASCAS)*, Puerto Vallarta, Mexico, pp. 1–4, 2018, doi: 10.1109/LASCAS.2018.8399972.
35. Agarwal, A. and Mishra, D., Altitude Optimization for DF Relaying Trajectory of UAV in Cooperative FANET. *GLOBECOM 2020 - 2020 IEEE Global Communications Conference*, Taipei, Taiwan, pp. 1–6, 2020, doi: 10.1109/GLOBECOM42002.2020.9322461.
36. Grasso, C., Raftopoulos, R., Schembra, G., Smart Zero-Touch Management of UAV-Based Edge Network. *IEEE Trans. Netw. Serv. Manage.*, 19, 4, 4350–4368, Dec. 2022, doi: 10.1109/TNSM.2022.3160858.
37. Bhardwaj, V. and Kaur, N., An efficient routing protocol for FANET based on hybrid optimization algorithm. *2020 International Conference on Intelligent*

Engineering and Management (ICIEM), London, UK, pp. 252–255, 2020, doi: 10.1109/ICIEM48762.2020.9160327.

38. Pandey, A., Shukla, P.K., Agrawal, R., Salp Swarm Optimization-Based Clustering Algorithm (SSOCA) in Adaptive FANET to Improve QoS for Disaster Response Operations. *Wirel. Pers. Commun.*, 126, 2801–2824, 2022, https://doi.org/10.1007/s11277-022-09842-4.
39. Kim, T., Lee, S., Kim, K.H., Jo, Y.-I., FANET Routing Protocol Analysis for Multi-UAV-Based Reconnaissance Mobility Models. *Drones*, 7, 161, 2023, https://doi.org/10.3390/drones7030161.
40. Nauman Bashir, M., Iqbal, S., Mohamad Yusof, K., Design Principles for Cooperative Relaying on UAVs-based FANET. *2022 Advances in Science and Engineering Technology International Conferences (ASET)*, Dubai, United Arab Emirates, pp. 1–6, 2022, doi: 10.1109/ASET53988.2022.9735058.
41. Chen, J., Xu, Y., Wu, Q., Zhang, Y., Chen, X., Qi, N., Interference-Aware Online Distributed Channel Selection for Multicluster FANET: A Potential Game Approach. *IEEE Trans. Veh. Technol.*, 68, 4, 3792–3804, April 2019, doi: 10.1109/TVT.2019.2902177.
42. Leonov, A.V. and Litvinov, G.A., About Applying AODV and OLSR Routing Protocols to Relaying Network Scenario in FANET with Mini-UAVs. *2018 XIV International Scientific-Technical Conference on Actual Problems of Electronics Instrument Engineering (APEIE)*, Novosibirsk, Russia, pp. 220–228, 2018, doi: 10.1109/APEIE.2018.8545755.
43. Tropea, M., Santamaria, A.F., Rango, F.D., Potrino, G., Reactive Flooding versus Link State Routing for FANET in Precision Agriculture. *2019 16th IEEE Annual Consumer Communications & Networking Conference (CCNC)*, Las Vegas, NV, USA, pp. 1–6, 2019, doi: 10.1109/CCNC.2019.8651744.
44. Leonov, A.V., Application of bee colony algorithm for FANET routing. *2016 17th International Conference of Young Specialists on Micro/Nanotechnologies and Electron Devices (EDM)*, Erlagol, Russia, pp. 124–132, 2016, doi: 10.1109/EDM.2016.7538709.
45. Bekmezci, I., Ermis, M., Kaplan, S., Connected multi UAV task planning for Flying *Ad Hoc* Networks. *2014 IEEE International Black Sea Conference on Communications and Networking (BlackSeaCom)*, Odessa, Ukraine, pp. 28–32, 2014, doi: 10.1109/BlackSeaCom.2014.6848998.
46. Faraci, G., Rizzo, S.A., Schembra, G., Green Edge Intelligence for Smart Management of a FANET in Disaster-Recovery Scenarios. *IEEE Trans. Veh. Technol.*, 72, 3, 3819–3831, March 2023, doi: 10.1109/TVT.2022.3217331.
47. Filho, J.A., Rosário, D., Rosário, D., *et al.*, Satisfactory video dissemination on FANETs based on an enhanced UAV relay placement service. *Ann. Telecommun.*, 73, 601–612, 2018, https://doi.org/10.1007/s12243-018-0658-z.

48. Joshi, A., Dhongdi, S., Kumar, S., Anupama, K.R., Simulation of Multi-UAV *Ad-Hoc* Network for Disaster Monitoring Applications. *2020 International Conference on Information Networking (ICOIN)*, Barcelona, Spain, pp. 690–695, 2020, doi: 10.1109/ICOIN48656.2020.9016543.
49. Lakhwani, K., Singh, T., Aruna, O., Multi-Layer UAV *Ad Hoc* Network Architecture, Protocol and Simulation, in: *Artificial Intelligent Techniques for Wireless Communication and Networking*, 2022, https://doi.org/10.1002/9781119821809.ch13.
50. Tropea, M., Fazio, P., De Rango, F., Cordeschi, N., A New FANET Simulator for Managing Drone Networks and Providing Dynamic Connectivity. *Electronics*, 9, 543, 2020, https://doi.org/10.3390/electronics9040543.
51. Kapil, J., Vivek, S.K., Vineet, S., Design of UAV Antennas and Challenges: Review. *Electr. Autom. Eng.*, 2, 1, 09–13, 2023, 10.46632/eae/2/1/2.
52. Zhu, L., Karim, M.M., Sharif, K., Xu, C., Li, F., Traffic Flow Optimization for UAVs in Multi-Layer Information-Centric Software-Defined FANET. *IEEE Trans. Veh. Technol.*, 72, 2, 2453–2467, Feb. 2023, doi: 10.1109/TVT.2022.3213040.
53. Grasso, C., Raftopoulos, R., Schembra, G., Slicing a FANET for heterogeneous delay-constrained applications. *Comput. Commun.*, 195, 362–375, 2022, ISSN 0140-3664, https://doi.org/10.1016/j.comcom.2022.08.024.
54. You, W., Dong, C., Wu, Q., Qu, Y., Wu, Y., He, R., Joint task scheduling, resource allocation, and UAV trajectory under clustering for FANETs. *China Commun.*, 19, 1, 104–118, Jan. 2022, doi: 10.23919/JCC.2022.01.009.
55. Mohammed Ahmed, S.B., Hussain, S.A., Latiff, L.A., Ahmad, N., Sam, S.M., Performance Evaluation of FANET Routing Protocols in Disaster Scenarios. *2021 IEEE Symposium On Future Telecommunication Technologies (SOFTT)*, Bandung, Indonesia, pp. 46–51, 2021, doi: 10.1109/SOFTT54252.2021.9673152.
56. Namdev, M., Goyal, S., Agarwal, R., An Optimized Communication Scheme for Energy Efficient and Secure Flying *Ad-hoc* Network (FANET). *Wirel. Pers. Commun.*, 120, 1291–1312, 2021, https://doi.org/10.1007/s11277-021-08515-y.
57. Cumino, P., Lobato Junior, W., Tavares, T., Santos, H., Rosário, D., Cerqueira, E., Villas, L.A., Gerla, M., Cooperative UAV Scheme for Enhancing Video Transmission and Global Network Energy Efficiency. *Sensors*, 18, 4155, 2018, https://doi.org/10.3390/s18124155.
58. Xia, X., Yan, Y., Li, Z., Meng, D., Xia, L., Chen, C., A Survey of UAV Clustering Algorithm. *2022 24th International Conference on Advanced Communication Technology (ICACT)*, PyeongChang Kwangwoon_Do, Korea, Republic of, pp. 191–196, 2022, doi: 10.23919/ICACT53585.2022.9728826.

59. Mukhopadhyay, A. and Ganguly, D., FANET based Emergency Healthcare Data Dissemination. *2020 Second International Conference on Inventive Research in Computing Applications (ICIRCA)*, Coimbatore, India, pp. 170–175, 2020, doi: 10.1109/ICIRCA48905.2020.9183223.
60. Surzhik, D.I., Vasilyev, G.S., Kuzichkin, O.R., Development of UAV trajectory approximation techniques for adaptive routing in FANET networks. *2020 7th International Conference on Control, Decision and Information Technologies (CoDIT)*, Prague, Czech Republic, pp. 1226–1230, 2020, doi: 10.1109/CoDIT49905.2020.9263944.

4

Quantum Cryptography for Secure FANET

Taskeen Zaidi[1] and Neha S.[2]*

[1]Department of CS&IT, Jain (Deemed-to-be University), Bangalore, Karnataka, India
[2]Master of Computer Application, Jain (Deemed-to-be University), Bangalore, Karnataka, India

Abstract

The security of a flying *ad hoc* network (FANET) is crucial because of its unique characteristics, such as its dynamic topology, high mobility, and decentralized network. The existing routing used by unmanned aerial vehicles (UAVs) comes with certain routing attacks, such as sinkhole attacks, dropping attacks, blackhole attacks, identity spoofing, data tampering, data leakage, and radio spoofing. The encrypting method used in UAV networks at present is based on mathematical complexity, which works quite well but can be easily decoded by quantum computers in no time. Quantum cryptography addresses the limitations of classical cryptographic methods. The protocol designed with a quantum mechanism based on the laws of physics will be a solution to many security threats for UAVs. The implementation of quantum cryptography requires specialized hardware that provides secure communication links, data protection, and detection of eavesdropping attempts. Post-quantum cryptography, the no-cloning theorem, and quantum key distribution (QKD) establish a shared secret key for both UAVs, which makes it impossible to be decoded by man-in-the-middle attacks.

Keywords: Security, cryptography, quantum computing, protocol

Abbreviations

AODV	*Ad-hoc* On-demand Distance Vector
ARP	Address Resolution Protocol
CA	Certificate authority

**Corresponding author*: nehakumar807323@gmail.com

Taskeen Zaidi, Adarsh Kumar, Saurav Mallik and Keshav Kaushik (eds.) Autonomous Flying Ad-Hoc Networks, (87–104) © 2025 Scrivener Publishing LLC

DNS	Domain Name Server
DOS	Denial of Service
DSR	Dynamic Source Routing protocol
ECC	Elliptic-curve cryptography
FANET	Flying *Ad-hoc* Network
GBS	Ground Base Station
HMACSHA1	Hash Message Authentication Code/Secure Hash Algorithmic
HTTP	Hypertext Transfer Protocol
IP	Internet Protocol
IOD	Internet of Drones
PKDC	Public Key Distribution Center
QPKE	Quantum Public Key Encryption
RSA	Rivest–Shamir–Adleman
SPOF	single point of failure
SQLi	Structured Query Language injection / SQL injection
SSL	Secure Sockets Layer
SYN	Synchronize
SYN-AKC	synchronize - acknowledgment
UAVs	Unmanned Aerial Vehicles

4.1 Introduction

Unmanned aerial vehicles (UAVs) are rapidly increasing. They have gained a lot of popularity in recent times. They are recognized for their wide range of applications, from military to commercial and civilian domains. The flying *ad hoc* network (FANET) has high mobility, low cost, and self-organizing features that are powerful enough to be incorporated into many domains. Hence, the technology must be reliable and secure. The identification of vulnerabilities and the development of solutions are essential for strengthening and securing FANETs.

4.2 Network Security Requirements

A FANET must meet the following parameters to satisfy basic network security requirements:

Authorization. The FANET must ensure nodes have access permissions for accessing information resources. Access control is intended to prevent unauthorized use of network services [1].

Integrity. The data transmitted from the source node must be received by the intended destination node without any malicious alteration or tampering. Only authorized nodes can modify and transmit messages [1].
Availability. Network services or resources must be available to legitimate users. It must sustain various attacks and tolerate failures in connectivity while still performing its expected services [1].
Confidentiality. Data transmitted from a source node must be received only by the intended destination node. An intruder must not be able to derive any useful information from the data. Confidentiality can be achieved *via* encryption [1].
Authenticity. Entities in the FANET must be able to identify each other. Identification of malicious or adversary entities helps prevent hindrances caused by them [1].
Non-repudiation. There should be no denying of any operation that was conducted. The sender of a message cannot deny having sent the message, and the recipient cannot deny having received the message. This is useful for the detection and isolation of compromised nodes [1].
Anonymity. Not publishing the identity of a node, in other words, keeping the identification information of a node private and sharing it only with authorized and intended users will prevent the identity of a node from reaching an adversary.

4.3 Security Threats

Unlike traditional wired or wireless networks, the FANET presents serious challenges in meeting security objectives. The wireless, infrastructure-free, and decentralized characteristics of FANET, as well as the lack of a centralized point of defense, make it more difficult to identify and respond to attacks. Barriers such as the weight constraints of UAVs result in inadequate physical protection and low processing and memory capacity, making it more difficult to implement cryptology methods that need a high level of memory and computation capability. In this section, we shall analyze the security threats in a FANET. Threats can be classified into active or passive, internal or external attacks.

Active Attack. An active attack attempts to disrupt the normal network functionality by altering or deleting data transmitted in the network. It can be an internal or external attack. Active internal attacks are performed by compromised nodes present in the network, and external attacks are performed by the adversary node that does not belong to the network. Active attacks include impersonation, modification, fabrication, and replication [3].

Passive Attack. A passive attack does not disrupt the network operations but sniffs the data being transmitted [3]. Confidentiality and privacy requirements get violated in a passive attack. A strong encryption mechanism must be implemented to avoid snooping or sniffing of data.

Internal and external attack. In external attacks, the adversary aims to disrupt the nodes and maliciously manipulate routing information. While in an internal attack, the adversary node enters the network by posing as an authorized node and engages in malicious actions.

4.3.1 Taxonomy of Security Threats/Attacks

4.3.1.1 Denial of Service Attack

In this attack, an adversary attempts to prevent legitimate UAVs from conducting network operations like packet forwarding and route discovery. The attacker aims to disrupt or decrease the performance of communication among the UAVs. A denial of service (DoS) attack can be carried out in many ways, such as by jamming at the physical layer, flooding packets to a victim UAV, or by an adversary taking part in the routing process and corrupting the normal functioning of the network. The DoS attacks mainly affect the FANET availability [1].

The following are the variants of a DoS attack:

Jamming. In a jamming attack, a jammer interferes with the radio signal between two nodes. The adversary generates strong signals to disrupt the communication of UAVs. It is a physical layer issue where the jamming of the signal causes messages to be corrupted or lost, leading to the FANET's disconnection [2].

Figure 4.1 demonstrates a basic model of a jamming attack. A jammer generates powerful radio signals (pulse and random noise are the most common types of jamming) that are strong enough to damage communication signals between UAV1, UV2, and UAV4.

SYN flooding attack. In a SYN attack, an adversary would send a huge number of SYN packets to a legitimate node and spoof the return addresses of the SYN packets. The victim node sends back SYN-ACK (acknowledge) packets to the attacker node, but the victim node does not receive any ACK packets back. Half-open connections and all pending connections would be stored on a fixed-size table. Eventually, the table gets overflown. This attack results in a situation where the victim will not be able to accept any other legitimate attempts to open a connection [2].

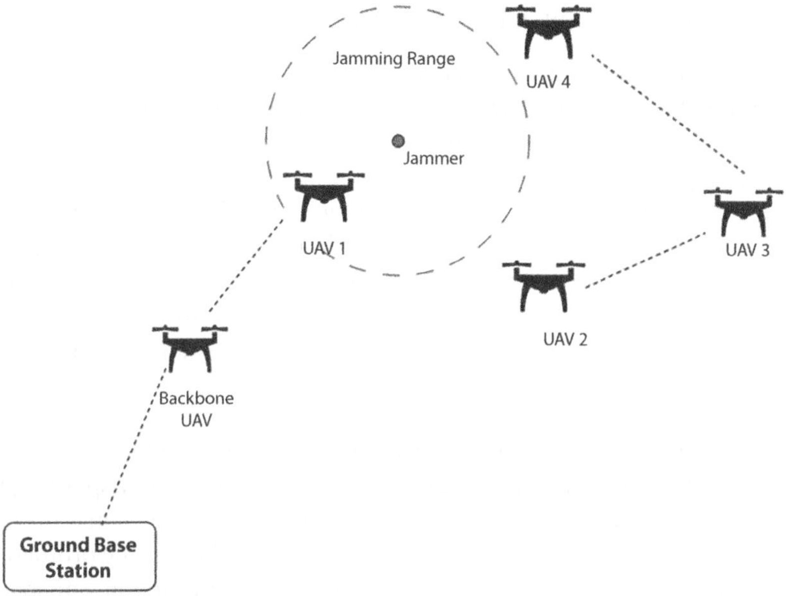

Figure 4.1 Jamming attack.

Repudiation attack. Repudiation attack refers to the denial of having participated in the communication. For example, a compromised node could delete or modify data and later deny having done so. This attack becomes a serious problem, as it would be difficult to track down the source and determine who is held accountable for the action [3].

Collision. Collision occurs when two or more nodes simultaneously send data in a half-duplex network environment. This results in an interruption of the flow of packets in the channel. For example, UAV1 and UAV2 are communicating with UAV3 at the same time. They are unaware of each other's simultaneous access to UAV3. Thus, a collision occurs when they both send data at the same time [4].

Distributed DoS. In this attack, several adversaries distribute themselves throughout the network and prevent legitimate UAVs from accessing the services offered by the network.

Resource consumption or sleep deprivation attack. The FANET is formed by multiple small UAVs that have limited memory, bandwidth, computation, and battery power capacity. In this attack, a malicious node tries to consume the resources of a legitimate node. An adversary could request the services of a certain node over again, which could be requests for routes, forwarding stale packets to nodes, etc. [3].

4.3.1.2 Modification and Fabrication Attacks

Under modification attacks, the adversary modifies the existing routing packets with malicious intent. Various methods can be deployed to alter the routing, such as changing the sequence number, the hop count, and altering the content of packets. The attacker spoofs their identity and creates some weakness in the network. Here, the integrity of the system is compromised [3, 5].

Under fabrication attacks, adversaries do not modify the existing routing information but create their own and damage the network's functionality. Variants of this attack affect the authenticity of the system. Rushing attacks, wormhole attacks, and grayhole attacks are instances of fabrication attacks [3, 5].

The following are subsets of modification or fabrication attacks:

Impersonation or identity spoofing. Impersonation or identity spoofing is a fabrication attack that affects the network's integrity. Impersonation happens due to a lack of authentication and integrity in the routing protocol. A compromised node impersonates a legitimate one, resulting in a misrepresentation of the network topology that causes network loops, partitioning, or bogus routing messages [3].

Blackhole attack. A malicious node falsely advertises itself as having a valid route (for instance, the shortest network destination route) to a legitimate destination UAV during the path-finding process or route updating message. The attacker then consumes the packets and never forwards them. Here, the adversary risks being monitored by its neighboring node and getting exposed. Thus, the attacker as shown in Figure 4.2 [4] modifies the received packets and selectively forwards the packets while leaving data from other nodes unaffected. The intention of the malicious node can be to hinder the path-finding process or intercept data packets being forwarded [2].

In Figure 4.3, a malicious node deceives UAV1 to transfer data. It then consumes the packets, modifies and selectively forwards packets to UAV3. UAV1, UAV2 and UAV3 are unaware of malicious nodes presents.

Wormhole attack. A wormhole, also referred to as a tunneling attack, occurs when one or more malicious nodes trick legitimate nodes in the network into sending packets to them at one location (point A) and tunnel them to another location in the network (point B). The packets are then retransmitted into the network (from point B). This tunnel between attacker UAVs is called a wormhole. An attacker records packets from a vital location in the network. When it is created in on-demand routing protocols such as the *ad hoc* on-demand distance vector (AODV) and dynamic

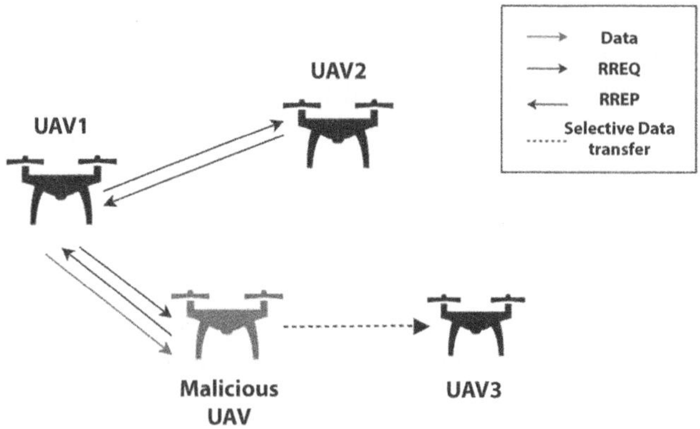

Figure 4.2 Blackhole attack.

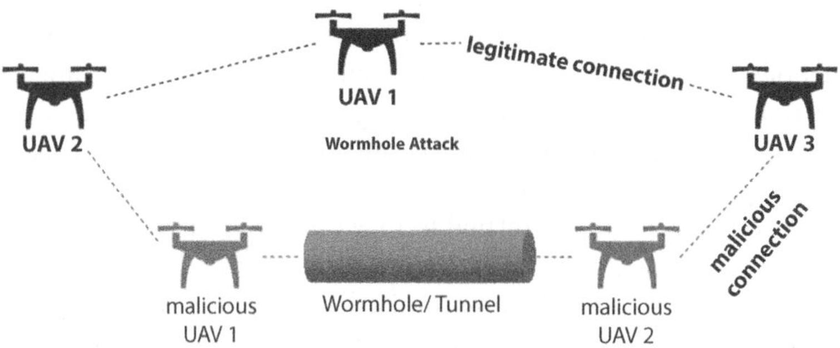

Figure 4.3 Wormhole attack.

source routing, it could prevent the route discovery process. If tunneling is done ethically, it can be used for efficient relaying of packets [4].

Man-in-the-middle attack. A man-in-the-middle attack is a type of attack where an adversary positions itself in between two legitimate UAVs. This can happen in two modes: active and passive. In passive mode, the adversary UAV can eavesdrop on communication, whereas in active mode, the UAV can delay, drop, or alter the messages transmitted. At the time of connection establishment, the adversary can capture sharing keys, do IP spoofing, ARP poisoning, DNS poisoning, ARP spoofing, DNS spoofing, SSL hijacking, and HTTP spoofing. In this attack, unauthorized UAVs can affect the confidentiality, integrity, and availability of FANET [4].

Byzantine attack. The Byzantine attack is performed by the nodes of the network. They pass all the authentication and verification processes. A compromised node or group of intermediate compromised nodes work in collision and carry out the attack of selectivity dropping, corrupting the packets, creating routing loops, forwarding the packets to a non-optimal path, etc. Byzantine behavior is hard to detect. The network seems to be operating normally from the perspective of other legitimate nodes, and Byzantine behavior is suspected [2].

4.3.1.3 Routing Attacks

Routing is the process of finding a path to establish communication between UAVs. Routing protocols in FANET are classically classified into proactive and reactive routing protocols. The attack on routing protocols can disrupt network operations, corrupt messages transmitted, block messages, create a blackhole in traffic, etc.

The attacks specific to routing protocols are listed below:

Routing table overflow. Routing information is stored in a routing table. Each node maintains a routing table containing up-to-date information about routes in the network. In a routing table overflow attack, an adversary floods incorrect routes to a legitimate node in the network. This causes the routing table of that victim node to overflow, which in turn prevents the addition of new entries related to new routes. Proactive (table-driven) routing protocols such as AODV are prone to this attack [3].

Route table poisoning. A compromised node pollutes routing updates or maliciously modifies and sends this route update packets to other legitimate nodes in the network. This attack causes severe damage to the network, which includes redirecting network traffic, causing routing loops, congestion, or an overall breakdown of the network [3].

Packet Replication. In this attack, an adversary replicates stale packets. This consumes additional resources, resulting in the depletion of memory and batteries on nodes, which affects the performance of nodes or causes congestion due to multiple copies of the same packet. Packet replication increases the delay, collision rates, and misdirection.

Routing cache poisoning. On-demand routing protocols maintain a route cache, which holds recently computed routes. An adversary can poison it to misdirect the network traffic, disrupting the network, injecting false route information, etc. [3].

Rushing attack. A rush occurs when the route discovery packets are being sent by attacks faster than other nodes. A dedicated channel or tunnel is created where packets can propagate faster than on a normal multi-hop route. By establishing faster packet forwarding, adversaries gain access rights to proceed with work and later replace legitimate routes for UAV service [3].

4.3.1.4 Other Attacks

Eavesdropping. Eavesdropping is also known as sniffing or snooping. It is a passive attack where the attacker passively listens to messages transmitted in the communication channel. This affects the privacy or confidentiality of a FANET. Eavesdropping is less harmful as the attacker does not modify or delete the messages. This attack mainly relies on an untrusted or unsecured network where the messages are not encrypted.

Malware injection/malware attack. Malware is an umbrella term for various malicious programs injected into FANET. Adversaries can harm the operation, network, and computer system, steal sensitive information, disrupt operations, and so on. Some common types of malicious code or malware are viruses, backdoor programs, worms, logic bombs, spyware, and Trojan horses.

GPS spoofing. Figure 4.4 illustrates [4] an adversary broadcasts false GPS to deceive the actual route. This can happen when the signal is not encrypted

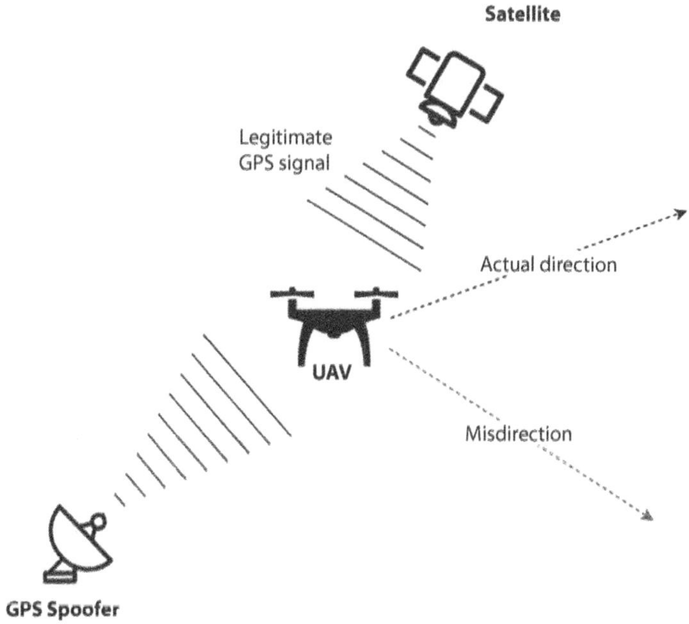

Figure 4.4 GPS spoofing.

or authenticated. A GPS simulator with a power amplifier and antenna is used to send false signal data. The UAV hijacking, operational failures, or crashes are caused by GPS spoofing.

Apart from attacks specifically imposed on the nodes and network, an adversary can attack the ground base station (GBS), client terminal, server, and storage. For example, the client terminal could be attacked to send malicious or fake commands, which affect FANET operation. Adversaries can perform data and/or device tampering to corrupt the terminal software or the hardware.

Attacks on GBS and backbone UAV cause single points of failure (SPOF) i.e., if the GBS or Backbone UAV fails, it results in the entire system failing. The redundancy of primary components such as having multiple GBS and backbone UAVs provides continuity operations in case of failure.

The SQL injection (SQLi)/command injection to the database causes manipulation of data and an execution of unauthorized commands.

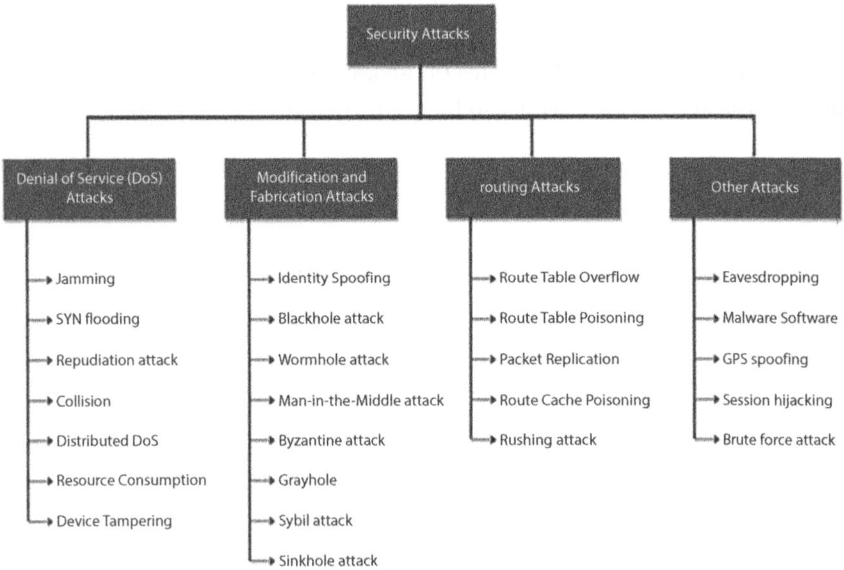

4.3.2 Summary

Attacks	**Concepts**	**Type [4]**	**Violation of security requirement [4]**
Blackhole	A malicious node participates in the network activities. Hinders the path finding process or intercept data packets being forwarded.	Active	Availability
Byzantine	One or more compromised nodes modify the packets, selectivity drops, or create routing loops.	Active	Availability, Data integrity
Collision	Two or more nodes simultaneously send data in a network environment, causing an interruption in the flow of packets.	Active	Availability
Distributed DoS	Adversaries distribute themselves in the network and disrupt the network functionality.	Active	Availability
Eavesdropping	An adversary snoops or sniffs the data without modifying the messages.	Passive	Authentication, authorization, confidentiality
GPS spoofing	An adversary broadcasts false GPS to deceive the actual route.	Active	Availability

(Continued)

(*Continued*)

Attacks	Concepts	Type [4]	Violation of security requirement [4]
Grayhole	A Grayhole attack has two phases. In the first phase, the AODV protocol is exploited. In the second phase, the node drops the intercepted packets [3].	Active	Availability
Identity spoofing	It impersonates a legitimate node, causing a misrepresentation of the network, resulting in network loops, partitioning, or bogus routing messages.	Active	Authentication, confidentiality, data integrity, non-repudiation
Jamming	The adversary generates strong signals to disrupt the communication in the network.	Active	Availability
Malware Software	Injecting a malicious program into the software, causing a disruption in the operation.	Active	Availability
Man-in-the-Middle	An adversary positioned between two legitimate nodes and modified the data transmitted.	Active	Confidentiality, data integrity
Packet replication	An adversary replicates the stale packets, causing the depletion of resources.	Active	Availability

(*Continued*)

(*Continued*)

Attacks	Concepts	Type [4]	Violation of security requirement [4]
Resource Consumption	A malicious entity tries to consume the resources of legitimate nodes.	Active	Availability
Repudiation	Denies having participated in the communication.	Active	Non-repudiation
Route Cache poisoning	In on-demand routing protocols, the adversary poisons the route cache.	Active	Integrity
Routing table overflow	The adversary floods incorrect routes to a legitimate node in the network, causing the routing table to overflow.	Active	Integrity, availability
Routing Table poisoning	A compromised node modifies routing updates, causing routing loops, congestion, or a breakdown of the network.	Active	Integrity, Anonymity, Confidentiality
Rushing	A dedicated channel is created where packets propagate faster than on a normal multi-hop route.	Active	Availability, Authorization

(*Continued*)

(*Continued*)

Attacks	**Concepts**	**Type [4]**	**Violation of security requirement [4]**
SYN flooding	An adversary would send a huge number of SYN packets to a legitimate node, and the node would not receive acknowledgments and create half-open connections and the table to overflow.	Active	Availability
Wormhole	One or more malicious nodes create a wormhole or tunnel, and packets are retransmitted into the network. An attacker records packets from a vital location.	Active	Integrity, confidentiality, availability

4.4 Quantum Cryptography

4.4.1 Quantum Cryptography Introduction

The authentication and communication schemes of IoD were designed based on RSA, ECC [6], HMACSHA1 algorithms, etc. The security of these schemes depends on the fact that no polynomial-time algorithm exists to solve prime factors of numbers [6]. But the quantum computers can decode these algorithms in polynomial time [6].

The quantum cryptography is based on the principles of quantum mechanism and not the "intractable" [7] nature of factoring (a very large number). The principles of quantum cryptography assure an unconditional security of information. In the experiment conducted in 2020 by Lui *et al.*, quantum drones were used as nodes for generating and measuring qubits [8]. Although this experiment was based on entanglement, it demonstrates the possibilities of quantum drones and quantum ground stations.

4.4.2 QPKE Based FANET Model (Based on the Encryption Model Introduced by Yuqi Wang)

Here there is an assumption that the servers in the control room, GBS, and the drones are equipped with quantum capabilities as shown in Figure 4.5. The proposed model is based on the QPKE (Quantum Public Key Encryption) scheme demonstrated by *Wang Y, She K*. QTF (Quantum Trapdoor Function) is used to produce quantum public key which takes a private key as an input.

A quantum trapdoor function is a one-way function that is easy to compute but hard to inverse [10]. In other words, it is difficult to find the input given the output. But the information "trapdoor" helps with efficient inversion. A QTF resembles the conventional trapdoor function but differs in two ways [10]:

1. The process generates a classical trapdoor and a quantum evaluation key.

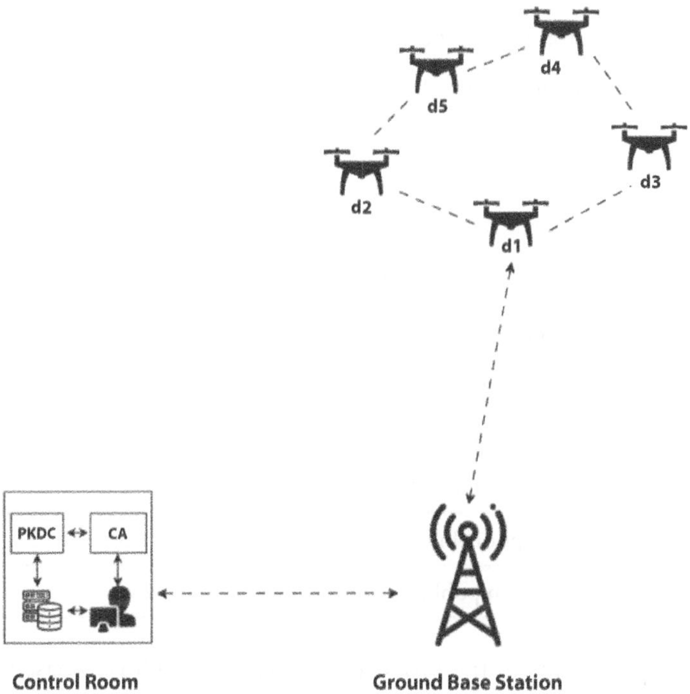

Figure 4.5 The control room is equipped with PKDC and CA for key management and certification. CA shares data block of quadruple to GBS and drones to establish session for quantum based secure communication.

2. Evaluation takes as input a classical string x, and the evaluation key, and outputs a quantum state.

Holevo X [9] quantity is used to protect communication between drones where published copies of each quantum public key corresponding to one private key must be restricted to an upper bound value, X.

A QPKE scheme for quantum messages is information-theoretic secure [9] ensuring unconditional security. The main requirement in this scheme is key management. The model requires trusted third parties [9] such as the PKDC (Public Key Distribution Center) for key distribution and management and CA (certificate authority) for identity authentication of the members of the FANET.

The following table contains the notations used in the algorithm:

Notations	Description
D	Drone
G	ground base station
ID_d	Identity of the drones
CA_{ID}	Drone's certificate
sk_{CA}	Drone's private key in corresponding to the drone's certification CA_{ID}
$GenAlg_{sk}$	Key generation algorithm
S_i	Private Key
F(.)	a trapdoor function
X	Holevo bound
(n,s_j)	Series of private key
D/E	Decryption/Encryption adapter
$\mid \Phi S_j \rangle$	Quantum public key of the given private key

The detailed steps of the proposed model are as follows (based on the encryption model introduced by Yuqi Wang):

1. CA load's identity authentication to drones (d) and d_n also obtain quadruple (ID, CA_{ID}, sk_{CA}, $GenAlg_{sk}$).

2. Drones execute $GenAlg_{sk}$ on their encryption and decryption adapter with their ID as input and generate a sequence of private key pairs (n,s_i) where $i \in [0, n-1]$, $n \in N$ and $n >> 1$, $s_i \in [0, n-1]$, n and s_i are unpredictable numbers.
3. Drones invoke and execute the quantum one way trapdoor function F(.) with the s_i as input data and generate equivalent Quantum public key for the private key in 1:1 relationship.
4. Drones register with PKDC for their quantum public keys equivalent to private key pair (n,s_i). If authenticated, PKDC accept the drone's quantum public key, registers and stores it. Else, PKDC denies the drone's request.
5. If a drone (for instance d_3) wants to communicate with d_1, d_3 sends request to CA center for ID_{d1}.

   Once ID_{d1} is obtained successfully

   And

   If d_3 also authenticated by CA center

   And

   If the value of d_3 counter is not more than Holevo χ

   Then PKDC will send quantum Public Key state $|\Phi s_i\rangle_{d1}$ to d_3 and keep record of it.

   Else PKDC will inform d_3 to upload its new Quantum Public Key and inform d3 to wait for a new Quantum Public Key.
6. The drone d_3 encrypts its data on its D/E adapter with Quantum Public Key state $|\Phi s_i\rangle_{d1}$ offered by PKDC and finally d_3 sends his encrypted data and its own signature offered by its certificate to d_1.
7. After receiving the data forwarded by the d_3, d_1 will request CA center to verify the identity of d_3. At the same, determine whether these encrypted quantum states are valid.

   If the cipher quantum state is valid d_1 will decrypt these cipher states and decreases d_1 counter by one.

   Else, d_1 shall notify d_3 to use the new quantum public key.

Centralization of public key distribution center and CA is one of the drawback of this model which can be improved using blockchain technology to decentralize the authentication and key distribution process.

4.5 Conclusion

FANET, being a highly recognized technology, has many schemes or protocols being developed to enhance security. This chapter gives an overview of security requirements, which act as primary parameter to analyze whether a FANET system is secure or not. It identifies the vulnerabilities of the flying *ad hoc* network and provides a detailed explanation and taxonomy of security threats and attacks.

The chapter later introduces quantum cryptography as a solution to the FANET security issues. A QPKE (Quantum Public Key Encryption) model that is based on the quantum trapdoor function has been proposed for securing FANET systems.

References

1. Wu, B., Chen, J., Wu, J., Cardei, M., A survey of attacks and countermeasures in mobile *ad hoc* networks, in: *Wireless Network Security*, pp. 103–35, 2007.
2. Biswas, K. and Ali, M.L., Security threats in mobile *ad hoc* network.
3. Jawandhiya, P.M., Ghonge, D.M., Ali, M.S., Deshpande, J.S., A survey of mobile *ad hoc* network attacks. *Int. J. Eng. Sci. Technol.*, 2, 9, 4063–71, 2010.
4. Tsao, K.Y., Girdler, T., Vassilakis, V.G., A survey of cyber security threats and solutions for UAV communications and flying *ad-hoc* networks. *Ad Hoc Netw.*, 133, 102894, 2022 Aug 1.
5. Korba, A.A., Nafaa, M., Salim, G., Survey of routing attacks and countermeasures in mobile *ad hoc* networks, in: *2013 UKSim 15th International Conference on Computer Modelling and Simulation*, IEEE, pp. 693–698, 2013 Apr 10.
6. Mishra, D., Singh, M., Reval, P., Pursharthi, K., Kumar, N., Barnawi, A., Rathore, R., Quantum-safe Secure and Authorized Communication Protocol for Internet of Drones. *IEEE Trans.Vehicular Tech.* 2023 Jul 6.
7. Aditya, J. and Rao, P.S., Quantum cryptography. *Proc. Comput. Soc.* India, 2005.
8. Abulkasim, H., Goncalves, B., Mashatan, A., Ghose, S., Authenticated Secure Quantum-Based Communication Scheme in Internet-of-Drones Deployment. *IEEE Access*, 10, 94963–72, 2022 Sep 6.
9. Wang, Y. and She, K., A practical quantum public-key encryption model, in: *2017 3rd International Conference on Information Management (ICIM)*, pp. 367–372, IEEE, 2017 Apr 21.
10. Coladangelo, A., Quantum trapdoor functions from classical one-way functions. *arXiv preprint arXiv:2302.12821*. 2023 Feb 24.

5

A Review of Various Routing Protocols for FANET

Nitya Nand Dwivedi

Shri Ramswaroop Memorial University, Barabanki, Uttar Pradesh, India

Abstract

Flying *ad hoc* networks (FANET) are products of cutting-edge research where many drones communicate. Drone use is increasing due to advances in sensors, processors, and applications. The FANETs are used when there are calamities, such as floods; they are also used for military purposes. The main problem of FANETs is the routing of these drones because high speed requires a dynamic routing process. In this chapter, we try to examine the proposed rules.

***Keywords*:** FANET routing, AeroRP, UAV

5.1 Introduction

There is a different routing scheme available for mobile *ad hoc* networks (MANETs). Different networks have individual requirements; hence, many routing algorithms have been designed to fulfill them. To decide which protocol must be used, many parameters must be investigated, such as bandwidth requirements, network architecture, convergence speed, and reliability [1]. The community of MANET researchers has suggested and implemented several routing protocols. There are two major approaches for routing path choice: one is the Bellman-Ford approach, the distance vector routing protocols, and for example, routing information protocol (RIP) while based on the Dijkstra approach and link state routing protocols like OSPF and optimized link state routing (OLSR). Both the Bellman-Ford approach and Dijkstra's approach are employed to find metrics (such

Email: nityananddwivedi29@gmail.com

Taskeen Zaidi, Adarsh Kumar, Saurav Mallik and Keshav Kaushik (eds.) Autonomous Flying Ad-Hoc Networks, (105–118) © 2025 Scrivener Publishing LLC

as distance, the cost of traversing a link) in routing protocols. Both methods use only hop counts as a metric of a link between two nodes. However, the bandwidth and delay can also be employed for metric calculation.

Distance vector routing protocols (DVRP): Every router of the network maintains routing update packets, which include an incline of all nodes in the network and their corresponding metric costs. The new packets are transmitted to all neighboring routers of the sending router. Similarly, every router receives these update packets from their neighbors based upon the fact that they update their routing tables.

Link-state routing protocols (LSRP): The router stores the map of the network topology in their memory. Then, the shortest path to every other router is calculated using Dijkstra's algorithm. Here also, the routing updates are periodic, such as when the topology changes. But the update packets, called link state packets (LSPs) are flooded into an entire network (unlike in the distance vector protocol where they are forwarded only to neighbors). The LSPs contain the ID of the source node, a sequence number (to decide which one is newer packets) and distance from the source to every neighbor. After collecting LSPs from all routers, routers create their routing table containing the shortest path to their respective destinations.

Routing in FANETS is very challenging since node mobility causes regular network topology updates. This requires flexible and robust routing solutions to keep up the existing routes and search for the new one when needed [2]. It is very difficult to find the shortest and most reliable path between places by going through nodes. When any node of the network moves, the established link may break. With a constantly updating topology, even maintaining a connection is quite difficult. Besides handling these regular topology changes, FANET routing protocols have to deal with other constraints, like a limited bandwidth, low energy consumption, and higher error probabilities, which may exist in a wireless environment. In addition to this, asymmetric links among mobile hosts and terrain conditions make routing more complicated. These networks need a responsive routing protocol that can find new valid routes as early as possible, whenever topology changes and the existing routes break. Since the employment of MANETs, many control policies in *ad hoc* networks have been considered and separated from the update process of generally used information as follows: demand-driven (reactive) protocols, hybrid protocols, table-driven (proactive) protocols, [3–5, 11].

Unmanned aerial vehicle (UAV) *ad hoc* networks are subclasses of vehicular *ad hoc* networks (VANETs) and MANETs. Routing in FANETs is quite different compared to traditional VANETs and MANETs, hence existing routing protocols will not be able to deal with such dynamic movements

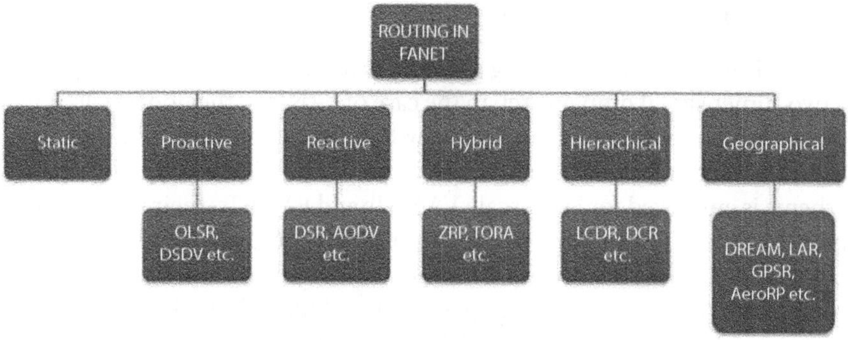

Figure 5.1 The deployment of a FANET routing protocol.

from aircraft. As a result, there is a major challenge of a link outage in these networks [22]. Current routing methods cannot cope with rapid changes in network quality, so they cannot be applied directly to FANETs [6]. The present research work in FANETs is based on either the modified version of the existing MANET routing approaches or the few novel approaches that have been presented. The FANETs, the routing approaches, can broadly be categorized as shown in Figure 5.1 [7].

5.2 Flying *Ad Hoc* Network Routing Protocol

Currently, there is no dedicated FANET root protocol in the industry, and the main protocol will continue to adopt the MANET protocol. The MANET route logs can be categorized into dissimilar categories, according to dissimilar configuration measures. It is divided into functional routes, active routes, and mixed routes, according to the route collection strategy.

5.2.1 Static Routing

In this routing in FANET, as shown in Figure 5.1, the aircraft is loaded with a routing table before starting any task or mission. The table cannot be modified at runtime. From here, this method is called a static method. Here, the connections between the aircraft remain fixed [8, 9].

The UAVs can communicate with other aircraft or ground stations to collect their data. In the event of failure of aircraft or any ground station, the routing table cannot be updated instantly. Instead, it will wait until task completion. Therefore, this approach is not illegal and is essential for the efficient operation of wireless topologies.

5.2.2 Hierarchical Routing

In a typical VANET, the routing protocol is represented as a routing plan in a two-dimensional domain. Yet, in the case of FANETs, the nodes will be flying in three-dimensional spaces along with different attributes like flight attitude, size, type of sensors and energy usage capabilities, etc. If a flat routing scheme is employed for large-scale VANETs and FANETs, their performance may degrade. To lower the possibility of this degradation and to improve network scalability, the use of hierarchical routing, also known as multi-level routing, can become one of the appropriate solutions [10, 11]. In this, aeronautical Ad-hoc networks (AANET) also known as Aeronautical Mobile Ad-hoc Networks (MANETs), are a type of network that permits aircraft to connect with each other wirelessly, forming a temporary network as they fly can be divided into several clusters, which consist of aircraft or UAVs. Each group has a representative called the group leader or cluster head (CH). The node will be responsible for group communication. The CH node will be in the direct communication array of each aircraft or UAV within the cluster. An aircraft within a cluster can communicate with other aircraft present in any other cluster, but only through their respective CH nodes. A CH is in direct connection with different ground stations, satellites, and aircraft, which are present in upper or lower layers. These multi-layers (or clustering) are better if aircraft are organized in different groups (busy in different assignments or missions), large numbers of aircraft are to be inter-connected and if the mission area is of large size.

5.2.2.1 Data-Centric Routing

Because aircraft communication is wireless, multicast is preferred over unicast between two communicating aircraft [12]. This method is used when distributing requests from multiple nodes using an algorithm. This promising routing approach can be used in FANETs [13, 14]. Here, the data requested and gathering is done based on data attributes and not by using the ID of the sender and receiver nodes. Each location or weather broadcast query (for example, "Take a video of region A if the change is greater than the specified percentage") collects specific text data about each location. Routing is completed and constructed on data content. Data collection algorithms can be used for effective data reporting. This method does three types of parsing: Space decoupling: Links do not need to contain information about each location or a unique ID. Time decoupling: The communicating nodes can communicate to the subscribers at the same time or later. Flow decoupling: Communication between two nodes can be accomplished in an asynchronous manner, i.e., data can be delivered constantly. This routing scheme is preferred for cluster-based structures in FANETs and when there are small numbers of aircraft on a predefined flight path. This helps in cooperation between distinct clusters.

5.2.2.2 Load, Carry, and Delivery Routing

Here, an aircraft is loaded by receiving data from any ground node (like getting any video image of the path to follow). Then, it carries these facts to any destination node (like any military team or a control station on the ground). Initially, the load-carry-and-delivery scheme was considered for single sources and single destinations. Multiple locations and multiple destinations can also be reached in the app because of the large separation between the two nodes [15–17]. This routing scheme is the right solution for latency insensitive and delay-tolerant applications (for examples of video images and secure data traffic). This routing solution can enhance the throughput with more security. However, if the distance between the source and the source is slightly greater, the delay will increase. This problem can be handled by developing multi-aircraft systems so that the distance between UAVs and transmission delay decreases.

5.2.3 Proactive Routing

Although routing is not required, this solution maintains updated routing information for most parts of the network. This requires transmitting connection information so that each receiver can calculate the route to the other part of the network. The routes can be computed using any standard algorithm like Dijkstra's. The link information can be disseminated by flooding the link information about any node with its neighboring nodes. In large networks, this flooding process can consume more bandwidth. So, this method is not suitable for mobile phones and large phones [18]. Here, a few proactive protocols which have been used before in various AANET implementations, are presented as follows:

5.2.3.1 Optimized Link State Routing

It is a dynamic routing protocol based on the use of multicast relay techniques to optimize link-state routing schemes. The routing process uses Hello messages and topology control (TC) messages to find neighbors [19]. Hello messages are used to determine directly connected neighboring nodes (i.e., one-hop neighbors). The TC report is used to generate basic topology information. This protocol can often be used in *ad hoc* networks with bandwidth and neighbors. The OLSR uses multi-point relay (MPR) technology to reduce management overhead. In case of motion and fault, the message about topology information for the whole network is changed periodically. The Hello message is received from a neighbor. These Hello messages are transmitted in the network only by selected MPR nodes. This way, overhead packets are minimized within the network, which may be created due to the forwarding of link state packets by every receiving node.

5.2.3.2 Destination-Sequenced Distance Vector (DSDV)

Another table generated by the Bellman-Ford algorithm is slightly modified to accommodate self-correlation. In the destination-sequenced distance vector (DSDV) shown in Figure 5.2, each node forms a table containing the arrangement numbers of all other nodes [20]. Whenever there is any topology update, the information is disseminated through an update mechanism (either a periodic or triggered update). These notifications may cause network connectivity issues. Hence, the concept of sequence numbers is employed in DSDV. This is done by the destination nodes. Newer routers with higher numbers are stored in the meeting table instead of routes with lower numbers. However, due to the periodical broadcasting of the routing table updates by each node, a large amount of overhead is generated.

5.2.4 Reactive Routing

This routing strategy is termed on-demand routing. The paths between two nodes is only created when they communicate. This is just to handle the high load of the operation. Here, the routing request is streamed from the source node to the network, and the response is answered by the receiving node. Communication begins when the forward request node receives the redirect response. Therefore, each node only manages the current usage. In

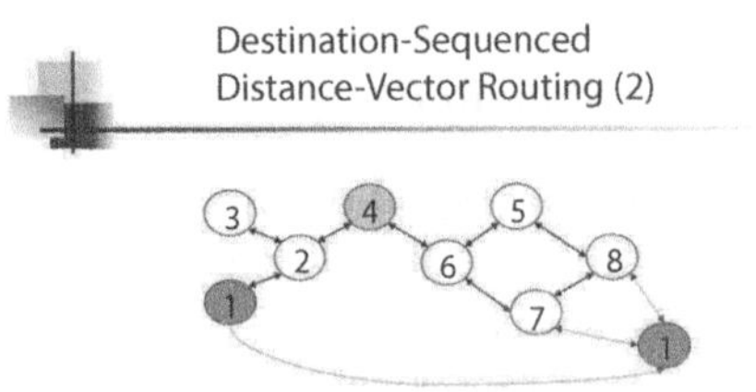

Figure 5.2 Destination-sequenced distance-vector routing.

this way, reactive routing is more bandwidth efficient. On the other hand, the discovery process takes a long time, so high latency can be one of the drawbacks.

5.2.4.1 *Dynamic Source Routing*

It is mostly considered for multi-hop mesh networking [21]. Here, any source node broadcasts a route demand message to the neighboring nodes present within its communication range. The source of the resource also adds an ID token to the message. In this way, the root node stores all the hop-off paths to the source node. If a source node cannot use the now-available route due to topology changes, then a route maintenance mechanism is used. In such belongings, the source node follows a new route to the destination. If the UAV has no way, it starts to look for a new way. This protocol was implemented for the UAV network by Brown *et al.* It is found that this issue may not be able to control the very high power of the FANETs.

5.2.4.2 *Ad Hoc On-Demand Distance Vector*

Other reactive processes have the same demand characteristics as dynamic source routing (DSR). In *ad hoc* on-demand distance vector (AODV), each of the meeting hosts, has access everywhere. In the case of DSR, each node has multiple entries for each location in its directory. In AODV, the sender and additional relay nodes store the succeeding hop information of the data sent at their conference, while the DSR packet stores the entire path between the sender and the destination domain [22]. The AODV routing process comprises three steps: route discovery, data packet transmission, and route maintenance. If one of the places has information to send, it first starts the search process to find the way to the place's location. After this, the data packets are forwarded over the determined route.

5.2.5 Hybrid Routing

It is a combination of passive and active routing methods. A hybrid solution can overcome the disadvantages of replication, such as high latency and extreme performance of the detection process appropriate for large networks. The network is separated into small areas where the local area communicates through the active layer and the regional communication is done using the reactive layer [23].

5.2.5.1 Zone Routing Protocol

In this routing protocol, every node has a separate zone having multiple nodes whose least distance to this node should be within the predefined radius. This way, the sectors of the neighboring nodes intersection. Routing within a domain is called intra-domain routing, where a proactive approach is used. If the destination and the destination are in the similar area, the destination can start sending packets directly [24]. If there is a need to communicate in the area, use the media as a good way to reduce the outcome of exploration. Here the answer is obtained only from the boundary of each field. Boundary nodes then continue to use the intermediate or connecting region as needed.

5.2.5.2 Temporarily Ordered Routing Algorithm (TORA)

It is a distributed protocol for multi-hop routing. Here, a router only maintains information of one-hop routers. It limits the amount of overhead (i.e., control messages) in a highly dynamic environment by limiting the responses to the topological updates [25]. Most of the process is based on rework, but it also has improvements with performance. It provides multiple paths between locations and the directed acyclic graph, which provides locations. In the event of a connection failure, a new route can be obtained quickly. It usually takes longer for TORA to reduce the check head. An individual node is assigned a value called "height" in the graph. No two nodes can have the same height. Since information flows like a liquid from the upper levels to the lower levels, its structure is not cyclical and irreversible. During the route discovery process, the querying node will get this height factor as a reply. In the process, every intermediate node will also update their TORA table with incoming heights and routes.

5.2.6 Geographic (or Position)-Based Routing

This routing method requires information about the location of the communication network. Generally, the global positioning system helps to calculate the physical location of individual nodes [26]. The various geographic forwarding decisions are breakdown into three types: nearest radius (MFR), nearest forward advance (NFP) and compass [27]. Here, the MFR sends the data sent in the largest transmission to the location of the place. The NFP transmits the packet to the source, closest to the source, and to the closest source to the destination. Compared to the MFR, it reduces the possibility of packet collisions as the resolution is shorter.

Compass guidance selects the node neighboring to the dotted line based on the location of the source and the destination's destination and continues. Some of the popular geo-routing protocols are discussed below:

5.2.6.1 DREAM (Temporarily Ordered Routing Algorithm)

At this point, the frequency of sharing location information between nodes is definite based on the distance between nodes and the movement speed of individual nodes. The further apart the nodes are, the fewer position updates need to be shared. In this way, DREAM optimizes the speed of producing control messages [28].

5.2.6.2 Location-Aided Routing

It is similarly based on the "wedge" thought used in DREAM. The wedge zone is the demand zone. This request field is used instead of a packet forwarding for requests [29, 30]. Location-aided routing (LAR) practices flooding. If a request to a route to nodes is confidential, the request zone does not reach the destination. In LAR, each node must know if it is in the request zone. There are two ways to tell if a node is in the desired area. In the first method, the sender sends a forwarding request containing the coordinates of the requested domain and a domain. The node receives the request and discards it if it is not in the rectangle, and forwards it if it is. In the second method, the requested area is not clearly specified, but data packets are sent based on the distance between source and destination.

5.2.6.3 Greedy Perimeter Stateless Routing

This protocol ahead of the data uses a greedy heuristic. The location information is shared with one-hop neighbors using a beaconing mechanism [31]. However, the data packets cannot always be sent using this greedy approach even when multiple paths exist. At this point, another approach called perimeter routing is used. Various techniques for traversing the perimeter of any planar graph are employed to forward the packets until the greedy approach can resume.

5.2.6.4 AeroRP

This protocol makes per-hop routing decisions in such a way that a packet is moved closer to the destination. For this, AeroRP uses a speed-based heuristic metric known as time-to-intercept (TTI). This protocol is used by the source node to get an idea about how soon different possible neighbors will be within the range of destination. The relative speed of the neighbor's ability to travel to the destination is the main factor in the TTI calculation. This speed is denoted as Sd. A positive Sd means the neighbor is moving towards the destination while a negative Sd means the neighbor is moving away.

5.2.7 Cross-Layer Routing

Sometimes, the layered network architectures do not apply to wireless communications [32]. This cross-layer routing method is suitable for improving performance in a wireless environment. It can be defined as a routing technique based on the desecration of the existing layered network architectures. There are many methods for designing cross-layer architecture [33]. Layers will be transferred to layers. Another cross-layer approach will provide interaction between non-adjacent layers. The FANET architecture is provided based on a cross-layer design approach [34]. Here, the collaboration of the first three layers of the open systems interconnection model is simplified. In this study, a new MAC protocol-Intelligent Media Access Control Protocol was issued. At the network layer, a directional optimized link state routing protocol is implemented in this system. The FANET performance can be upgraded by using this cross-layer design. Another algorithm called the meshed-tree algorithm, which is established on the directional antennas, was proposed by Huba and Shenoy [35]. Here, the grouping and scheduling at the MAC layer are integrated with the routing process at the network layer. The scheduling is based on the TDMA technique. TDMA represents multiple accesses in a time division. It is a digital wireless telephone transmission technology that divides channels into time windows, allowing multiple users to share the same frequency channel by sending them to different time windows. This approach improves packet delivery speed and end-to-end latency.

5.3 Conclusion

The FANET's control systems are now attracting more attention from researchers due to their different features. Here, we try to mention almost maximum number of methods used by the FANET because there is still more work to be done in this area and more routing protocols to be found.

References

1. Maxa, J.-A., Mahmoud, M.-S.B., Larrieu, N., Survey on uaanet routing protocols and network security challenges. *Ad Hoc Sens. Wirel. Netw.*, 2017.
2. Royer, E.M. and Toh, C.-K., A review of current routing protocols for ad hoc mobile wireless networks. *IEEE Pers. Commun.*, 6, 2, 46–55, April 1999.
3. Broch, J., Maltz, D.A., Johnson, D.B., Hu, Y.-C., Jetcheva, J., A performance comparison of multi-hop wireless ad hoc network routing protocols, in: *Proceedings of the 4th annual ACM/IEEE international conference on Mobile computing and networking*, ACM, pp. 85–97, 1998.
4. Ismail, R., Zulkifli, C.Z., Samsudin, K., Routing protocols for mobile *ad-hoc* network: A qualitative comparative analysis. *J. Teknol.*, 78, 8, 1–10, 2016.
5. Nayyar, A., Flying adhoc network (FANETs): Simulation based performance comparison of routing protocols: AODV, DSDV, DSR, OLSR, AOMDV and HWMP. *International Conference on Advances in Big Data, Computing and Data Communication Systems (icABCD)*, 08 2018, pp. 1–9.
6. Jiang, J. and Han, G., Routing protocols for unmanned aerial vehicles. *IEEE Commun. Mag.*, 56, 1, 58–63, Jan 2018.
7. Tareque, M.H., Hossain, M.S., Atiquzzaman, M., On the routing in flying ad hoc networks, in: *2015 Federated Conference on Computer Science and Information Systems (FedCSIS)*, Sep. 2015, pp. 1–9.
8. Franchi, A., Secchi, C., Ryll, M., Bulthoff, H.H., Giordano, P.R., Shared control : Balancing autonomy and human assistance with a group of quadrotor UAVs. *IEEE Rob. Autom. Mag.*, 19, 3, 57–68, Sep. 2012.
9. Lamont, G.B., Slear, J.N., Melendez, K., UAV swarm mission planning and routing using multi-objective evolutionary algorithms, in: *2007 IEEE Symposium on Computational Intelligence in Multi-Criteria Decision-Making*, April 2007, pp. 10–20.
10. Sun, Z., Wang, P., Vuran, M.C., Al-Rodhaan, M.A., Al- Dhelaan, A.M., Akyildiz Bordersense, I.F., Border patrol through advanced wireless sensor networks. *Ad Hoc Networks*, 9, 3, 468–477, 2011.
11. Ko, J., Mahajan, A., Sengupta, R., A network-centric UAV organization for search and pursuit operations, in: *Proceedings, IEEE Aerospace Conference*, March 2002, vol. 6, pp. 6–6.
12. de Jong, E., Flexible data-centric UAV platform eases mission adaptation, white paper, available online, 2011.
13. Koller, A. and Johnson, E., Design, implementation, and integration of a publish/subscribe-like multi-UAV communication architecture, in: *AIAA Modeling and Simulation Technologies Conference and Exhibit*, June 2012, p. 6309.
14. Cheng, C., Hsiao, P., Kung, H.T., Vlah, D., Maximizing throughput of UAV relaying networks with the load-carry-and-deliver paradigm, in: *2007 IEEE Wireless Communications and Networking Conference*, March 2007, pp. 4417–4424.

15. Le, M., Park, J., Gerla, M., UAV assisted disruption tolerant routing, in: *MILCOM 2006 - 2006 IEEE Military Communications conference*, Oct 2006, pp. 1–5.
16. Jonson, T., Pezeshki, J., Chao, V., Smith, K., Fazio, J., Application of delay tolerant networking (DTN) in airborne networks, in: *MILCOM 2008 - 2008 IEEE Military Communications Conference*, Nov 2008, pp. 1–7.
17. Huba, W. and Shenoy, N., A MANET architecture for airborne networks with directional antennas. *Int. J. Adv. Netw. Serv.*, 333–3455, 3 & 4, 2012.
18. Clausen, T. and Jacquet, P., Optimized link state routing protocol (OLSR), Technical report, 2003.
19. Perkins, C.E. and Bhagwat, P., Highly dynamic destination-sequenced distance-vector routing (DSDV) for mobile computers, in: *ACM SIGCOMM computer communication review*, vol. 24, pp. 234–244, ACM, 1994.
20. Johnson, D.B., Maltz, D.A. (1996). Dynamic Source Routing in Ad Hoc Wireless Networks. In: Imielinski, T., Korth, H.F. (eds) Mobile Computing. The Kluwer International Series in Engineering and Computer Science, vol 353. Springer, Boston, MA. https://doi.org/10.1007/978-0-585-29603-6_5
21. Perkins, C., Belding-Royer, E., Das, S., Ad hoc on-demand distance vector (AODV) routing, Technical report, 2003.
22. Wang, L., Olariu, S. (2005). Hybrid Routing Protocols for Mobile Ad-hoc Networks. In: Cardei, M., Cardei, I., Du, DZ. (eds) Resource Management in Wireless Networking. *Network Theory and Applications*, vol 16. Springer, Boston, MA. https://doi.org/10.1007/0-387-23808-5_17
23. Haas, Z.J., Pearlman, M.R., Samar, P., The zone routing protocol (ZRP) for ad hoc networks, Internet Engineering Task Force (IETF), 2002.
24. Park, V.D. and Corson, M.S., A highly adaptive distributed routing algorithm for mobile wireless networks, in: *Proceedings of INFOCOM '97*, April 1997, vol. 3, pp. 1405–1413.
25. Oubbati, O.S., Lakas, A., Zhou, F., Güneş, M., Yagoubi, M.B., A survey on position-based routing protocols for flying ad hoc networks (FANETs). *Veh. Commun.*, 10, 29–56, 2017.
26. Peters, K., Jabbar, A., Çetinkaya, E.K., Sterbenz, J.P.G., A geographical routing protocol for highly-dynamic aeronautical networks, in: *2011 IEEE Wireless Communications and Networking Conference*, March 2011, pp. 492–497.
27. Woodward, B.A., A distance routing effect algorithm for mobility DREAM, in: *Proc. of the 4th annual ACM/IEEE international conference on Mobile computing and networking*, October 1998, pp. 76–84.
28. Su, W., Lee, S.-J., Gerla, M., Mobility prediction and routing in ad hoc wireless networks. *Int. J. Netw. Manage.*, 11, 1, 3–30, 2001.
29. Ko, Y.-B. and Vaidya, N.H., Location-Aided Routing (LAR) in mobile *ad hoc* networks. *Wireless Netw.*, 6, 4, 307–321, 2000.

30. Karp, B. and Kung, H.-T., GPSR: Greedy perimeter stateless routing for wireless networks, in: *Proceedings of the 6th annual international conference on Mobile computing and networking*, ACM, pp. 243–254, 2000.
31. Srivastava, V. and Motani, M., Cross-layer: A survey and the road ahead. *IEEE Commun. Mag.*, 43, 12, 112–119, Dec 2005.
32. Foukalas, F., Gazis, V., Alonistioti, N., Cross-layer proposals for wireless mobile networks: A survey and taxonomy. *IEEE Commun. Surv. Tutor.*, 10, 1, 70–85, First 2008.
33. Alshbatat, A.I. and Dong, L., Cross layer for mobile ad-hoc unmanned aerial vehicle communication networks, in: *2010 International Conference on Networking, Sensing and Control (ICNSC)*, April 2010, pp. 331–336.
34. Huba, W. and Shenoy, N., Airborne surveillance networks with directional antennas, in: *ICNS 2012, The Eighth International Conference on Networking and Services*, pp. 1–7, 2012.
35. Alshabtat, A.I., Dong, L., Li, J., Yang, F., Low latency routing algorithm for unmanned aerial vehicles *ad-hoc* networks. *Int. J. Electr. Comput. Eng.*, 6, 1, 48–54, 2010.

6

The Integration of the Internet of Things in FANET

Ankur Chaudhary[1*], Neetu Faujdar[2] and Ritesh Rastogi[1]

[1]Department of Information Technology, Noida Institute of Engineering & Technology, Greater Noida, India
[2]Department of Computer Engineering and Application, GLA University, Mathura, India

Abstract
The integration of the Internet of Things (IoT) with flying *ad hoc* networks (FANET) holds considerable potential to revolutionize numerous industries. The FANET is a network of unmanned aerial vehicles (UAVs) or autonomous drones that communicate with one another to carry out a certain task or goal. Internet of Things technology can be integrated with FANET to provide a number of benefits, such as data gathering and sharing, independent tasks, network availability, traffic control, information investigation, and bits of knowledge. However, it is essential to take into consideration the difficulties that come with integrating the IoT into FANET. These difficulties include managing power consumption, ensuring secure communication, addressing privacy concerns, and maintaining robust network connectivity in dynamic environments. The FANET is used in a wide range of fields and sectors, such as observation and security, search and salvage, management of disasters, accuracy farming, connectivity and communication, and ecological checking. It is important to consider the difficulties that come with integrating the IoT into FANET, such as managing power consumption, ensuring secure communication, and addressing privacy concerns.

***Keywords*:** UAV, FANET, MANET, IDS, cryptography, routing, topology, attack

**Corresponding author*: ankurchaudhary849@gmail.com; ORCID: https://orcid.org/0000-0002-8756-9157
Neetu Faujdar: ORCID: https://orcid.org/0000-0002-1482-9755
Ritesh Rastogi: ORCID: https://orcid.org/0000-0003-0427-5290

Taskeen Zaidi, Adarsh Kumar, Saurav Mallik and Keshav Kaushik (eds.) Autonomous Flying Ad-Hoc Networks, (119–148) © 2025 Scrivener Publishing LLC

6.1 Introduction

For aerial vehicles and unmanned systems, the integration of the Internet of Things (IoT) with flying *ad hoc* networks (FANET) holds considerable potential to revolutionize numerous industries. A network of unmanned aerial vehicles (UAVs) or autonomous drones that communicate with one another to carry out a certain task or goal is referred to as a FANET.

6.1.1 Integration of IoT Technology with a FANET

Internet of Things technology can be integrated with a FANET to provide a number of benefits, including:

a) **Data gathering and sharing:** Drones can be equipped with IoT components like sensors and actuators to gather environmental data in real-time. These sensors can keep an eye on a number of variables, including temperature, humidity, air quality, and even the detection of particular objects. The networked drones can then exchange this data to facilitate group decision-making and improve situational awareness.
b) **Independent tasks:** Based on the data collected, an IoT-enabled FANET can make intelligent decisions by utilizing artificial intelligence (AI) and machine learning algorithms. Drones can independently analyze the data, alter their flight paths, and alter their behavior. Applications like environmental monitoring, search and rescue, and surveillance all benefit greatly from this capability.
c) **Network availability:** With the IoT, drones and other systems or devices on the ground can connect in real time. Robots can go about as information transfers, furnishing network admittance to regions with restricted or no current foundation. This network can uphold applications like remote detecting, debacle reaction, or accuracy farming.
d) **Traffic control:** Coordinating IoT into a FANET empowers effective traffic to the executives of aeronautical vehicles. To ensure safe and collision-free operations, each drone can communicate its position, speed, and trajectory to other drones in the network. Increased scalability and efficiency

are brought about by coordinated flight plans and route optimization made possible by this level of communication.

e) **Information investigation and bits of knowledge:** The IoT-enabled FANET's data can be processed and analyzed to provide useful insights. The ability of machine learning algorithms to identify patterns, trends, and correlations makes it possible to make better decisions and optimize operations. Applications like disaster management, infrastructure monitoring, and urban planning can all benefit from these insights.

However, it is essential to take into consideration the difficulties that come with integrating the IoT into a FANET. These difficulties include managing power consumption, ensuring secure communication, addressing privacy concerns, and maintaining robust network connectivity in dynamic environments. Conquering these difficulties will be critical for the fruitful execution and far-reaching reception of the IoT in a FANET.

6.1.2 Overview of FANET and Its Applications

The FANETs are self-organizing networks of UAVs communicating with one another to accomplish a common objective. Without the need for a centralized infrastructure or a pre-existing communication network, these networks can function independently. The FANETs are intended to empower joint effort and coordination among drones, permitting them to productively perform complex assignments [1].

A FANET is used in a wide range of fields and sectors. A few striking models include:

a) **Observation and security:** A FANET can be used for observation and security purposes. Drones outfitted with cameras and sensors can screen enormous regions, distinguish irregularities, and communicate constant video feeds to ground stations or control focuses. Security and surveillance operations can benefit from enhanced coverage, adaptability, and quick response capabilities thanks to a FANET.

b) **Search and salvage:** A FANET assumes an essential part in search-and-salvage missions, especially in remote or difficult-to-reach regions. Drones can be used to send vital information to rescue teams, survey disaster-ravaged areas, and find missing people. A FANET's autonomy makes it possible to conduct effective search patterns, speed up response times, and increase the likelihood of successful rescues.

c) **Management of disasters:** FANET can help with calamity the board and moderation endeavors. Drones outfitted with sensors and specialized gadgets can evaluate the degree of harm, recognize possible dangers, and help with planning salvage and alleviation activities. A FANET makes it possible to quickly collect, analyze, and communicate data, facilitating informed decision-making in times of emergencies.

d) **Accuracy farming:** A FANET has applications in the agrarian area, empowering accuracy and cultivating strategies. Drones with sensors and cameras can optimize irrigation and fertilizer application, monitor crop health, and detect diseases or pests. A FANET aids information assortment, examination, and navigation, prompting further developed crop yields and assets for the executives.

e) **Connectivity and communication:** A FANET can give an impermanent or crisis correspondence foundation in regions with restricted or disturbed networks. Drones outfitted with specialized gadgets can go about as elevated base stations, handing off signals between ground gadgets or laying out impromptu organizations. A FANET makes it easier for disaster-stricken areas, events that need temporary communication networks, and remote areas to connect.

f) **Ecological checking:** A FANET empowers proficient natural checking and research. Drones outfitted with sensors can gather information on air quality, water assets, natural living spaces, and biological changes. A FANET considers wide-region inclusion, constant information assortment, and versatile inspecting, supporting natural preservation and exploration drives.

These are only a couple of instances of the uses of FANET. As innovation advances and guidelines develop, we can anticipate that a FANET should track down new applications and add to different fields, including transportation, coordinated factors, foundation investigation, and amusement, among others. Figure 6.1 shows development and deployment phase of FANET.

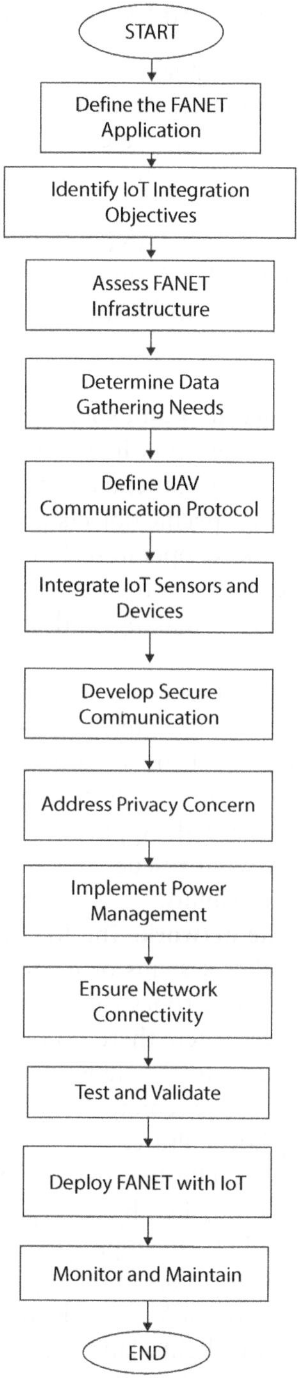

Figure 6.1 Development phases.

6.1.3 Introduction of IoT and Its Relevance in FANET

The physical devices, vehicles, appliances, and other objects that are embedded with sensors, software, and connectivity capabilities are referred to as the IoT. These devices can communicate and interact with each other, as well as with humans, by collecting and exchanging data over the internet [2].

The pertinence of the IoT in a FANET lies in its capacity to improve the abilities and usefulness of UAVs and empower consistent availability among them. By incorporating IoT advancements into a FANET, a few benefits can be understood:

a) **Information assortment and sharing:** Drones can be outfitted with IoT components like actuators and sensors to collect real-time environmental data. These sensors are capable of detecting specific objects and monitoring a variety of parameters like temperature, humidity, and air quality. The networked drones can then share this data, facilitating collaborative decision-making and increasing situational awareness.
b) **Independent tasks:** Based on the data collected, an IoT-enabled FANET can make intelligent decisions using AI and machine learning algorithms. Robots can independently investigate the data, adjust their flight paths, and change their behavior as needed. Applications like environmental monitoring, search and rescue, and surveillance all benefit greatly from this capability.
c) **Connection to the network:** The IoT can make it possible for drones and other systems or devices on the ground to connect in real-time. Drones can serve as data relays, giving people in places where there is not much infrastructure or none at all access to the network. This availability can uphold applications like remote detecting, catastrophe reaction, or accuracy horticulture.
d) **Traffic control:** The effective management of aerial vehicle traffic is made possible by incorporating the IoT into a FANET. Each robot can impart its situation, speed, and direction to different robots in the organization, guaranteeing safe and crash-free tasks. This degree of correspondence considers composed flight plans and course improvement, bringing about upgraded effectiveness and adaptability.

e) **Upkeep and armada the executives:** The IoT mix can work with prescient drone support. The UAVs' health and performance can be monitored in real-time by sensors, spotting anomalies or potential failures. This data can be utilized to plan support tasks proactively, decreasing margin time and working on the general unwavering quality of the armada.
f) **Insights and analysis from data:** The information gathered by an IoT-empowered FANET can be handled and examined to determine significant bits of knowledge. The ability of machine learning algorithms to identify patterns, trends, and correlations makes it possible to make better decisions and optimize operations. Applications like disaster management, infrastructure monitoring, and urban planning can all benefit from these insights.

The networked drones can operate more effectively, autonomously, and intelligently by utilizing a FANET's IoT capabilities. This combination upgrades their capacity to gather and share information, speak with different gadgets, improve tasks, and add to a great many applications across enterprises [3].

6.1.4 Importance of Integrating IoT with FANET for Enhanced Capabilities

The reconciliation of the IoT with a FANET brings a few significant advantages and upgrades the capacities of UAVs in different ways:

a) **Improved situational mindfulness:** IoT coordination permits drones in the FANET to accumulate constant information from the climate through sensors. This information can incorporate data like temperature, moistness, air quality, and other applicable boundaries. Drones can make well-informed decisions, adjust to changing conditions, and effectively respond to their operating environment if they have access to this information. The overall performance and effectiveness of FANET operations are enhanced by this enhanced situational awareness
b) **Making decisions together:** Data can be shared between drones that are connected through an IoT-enabled FANET. This cooperative climate permits robots to share sensor information, flight plans, and other significant data, prompting

better navigation. Robots can speak with one another to streamline flight paths, stay away from crashes, and direct their activities for mission achievement. The capacity to trade continuous information improves the general insight and productivity of the FANET.

c) **Operations on their own:** The mix of IoT and FANET empowers robots to work independently and pursue keen choices in light of gathered information. Drones are capable of adapting their behavior in response to patterns found in sensor data. Drones, for instance, have the ability to dynamically adjust their flight paths on their own to meet shifting mission requirements or avoid obstacles. This degree of independence decreases the requirement for human mediation and takes into account more proficient and versatile tasks.

d) **Consistent network:** The IoT is incorporated with consistent availability among drones and different gadgets or frameworks. Robots can go about as correspondence transfers, furnishing network admittance to regions with restricted foundation or availability. Remote control, real-time data transmission, and the exchange of crucial information between drones and ground stations or other IoT devices are all made possible by this connectivity. The FANET's overall efficiency and coverage are enhanced by the seamless connectivity.

e) **Information driven bits of knowledge and examination:** IoT reconciliation empowers the enormous volumes of information from drones inside the FANET. This information can be handled and broken down using progressed examination strategies, for example, AI and man-made consciousness. Decision-making, operational optimization, and performance evaluation can all benefit from the insights gleaned from the data. Analytics, for instance, can assist in locating anomalies, trends, or patterns in sensor data, which can improve resource allocation, predictive maintenance, and overall operational efficiency.

f) **Versatility and adaptability:** IoT integration enhances a FANET's scalability and adaptability. As the quantity of IoT-empowered drones builds up, the organization can adjust and self-sort out to oblige the extending network size. In addition, the adaptability of IoT devices enables the easy integration and deployment of new communication modules, actuators, or sensors, allowing for adaptation to specific operational scenarios or mission requirements.

Generally, incorporating the IoT with a FANET gives upgraded capacities, including advanced situational mindfulness, cooperative direction, independent activities, consistent network, information-driven bits of knowledge, and adaptability. These advantages add to more productive, savvy, and compelling tasks for many applications, making IoT mix urgent for opening the maximum capacity of a FANET [4].

6.2 Fundamentals of FANETs

Flying *ad hoc* networks are self-organizing networks of UAVs communicating with one another to accomplish a common objective. They work in a decentralized way, without depending on a proper foundation or brought together control. The fundamentals of FANET are as follows:

a) **Unmanned aerial vehicle nodes:** A FANET comprises of different UAV hubs that are fit for an independent flight. Depending on the application, these UAVs can differ in size, capabilities, and payload. With sensors, processors, actuators, and communication tools, each UAV in the FANET functions as a node.
b) **Impromptu systems administration:** A FANET uses impromptu systems administration standards, where UAVs structure a powerful organization on the fly. Utilizing wireless communication protocols like Wi-Fi, Bluetooth, or specialized mesh networking protocols, the UAV nodes communicate with one another. This correspondence empowers the trading of control messages, information, and other significant data.
c) **Self-organization:** FANET hubs independently self-sort out into an organization without depending on brought-together coordination. The UAVs find and interface with neighboring hubs, lay out correspondence, and progressively adjust the organization geography in view of the development and accessibility of hubs. A FANET is able to adjust to shifting circumstances, improve routing, and maintain network connectivity thanks to this self-organization.

d) **Versatility and dynamic geography:** A FANET manages profoundly unique and portable hubs. The UAVs can move in a three-layer space, change elevation, and change their flight ways given the mission prerequisites or ecological circumstances. Maintaining connectivity, managing the network topology, and ensuring that data and control messages are efficiently routed are all issues that come with FANET's dynamic nature.
e) **Steering and information trade:** A FANET requires proficient directing conventions to empower correspondence between UAV hubs. Directing calculations in a FANET are intended to think about the unique geography, restricted power assets, and irregular network. Considerations like link quality, node mobility, and energy efficiency are taken into account by these algorithms when determining the best routes for data exchange.
f) **Mission-explicit objectives:** Depending on the application, FANETs are used for specific missions or goals. The objectives can go from observation and surveillance to calamity reaction, ecological checking, or accuracy agribusiness. The mission-specific requirements and constraints, such as real-time data collection, collaborative decision-making, or adaptive flight paths, are supported by the FANET architecture and protocols.
g) **Security and wellbeing:** A FANET's safety and security are crucial features. As UAVs work in shared airspace, it is fundamental to guarantee secure correspondence, forestall unapproved access, and safeguard against digital dangers. Security measures incorporate impact aversion, consistence with aeronautics guidelines, and adherence to flight limitations and airspace rules.

By making intelligent, autonomous, and collaborative aerial operations possible, FANETs have the potential to transform different industries. Be that as it may, difficulties in the network and the board, and challenges in energy productivity, versatility, and mixing with existing avionics frameworks should be tended to for a more extensive reception and effective arrangement of FANETs in real situations.

6.2.1 Explanation of FANET Architecture and Operation

A Fundamentals of FANET Figure 6.2 shows UAVs that speak with one another in a distributed style. The FANETs are normally utilized for applications where a concentrated organization framework is inaccessible or not doable, like in a debacle or military tasks. The engineering of a FANET can change contingent upon the application. However, there are two primary sorts of FANET structures: unified and decentralized. Brought together, FANET engineering utilizes a ground control station (GCS) as the focal hub of the organization. All information correspondence between UAVs is steered through the GCS. This kind of engineering is easy to carry out and make due, yet it may be a bottleneck on the off-chance that the GCS is overburdened. However, it is more adaptable, and open-minded than a unified FANET design. The activity of a FANET depends on the standards of impromptu systems administration. Networks that are not preconfigured or managed by a central authority are known as *ad hoc* networks. The UAVs in a FANET are self-put together and structure an organization powerfully. They use steering conventions to trade directing data and to course information parcels between UAVs. Figure 6.3 shows the various types of FANET.

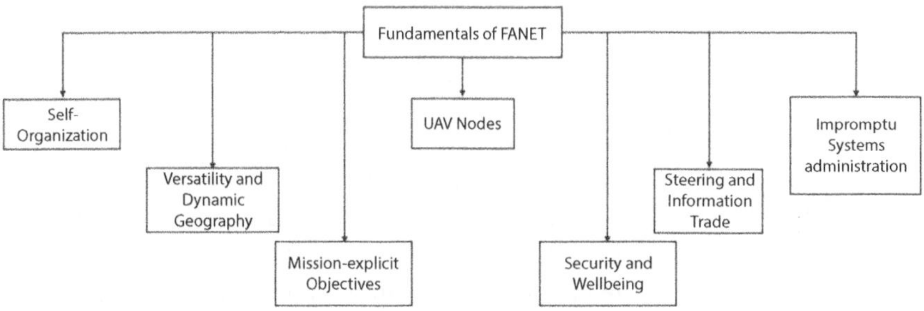

Figure 6.2 Fundamentals of FANET.

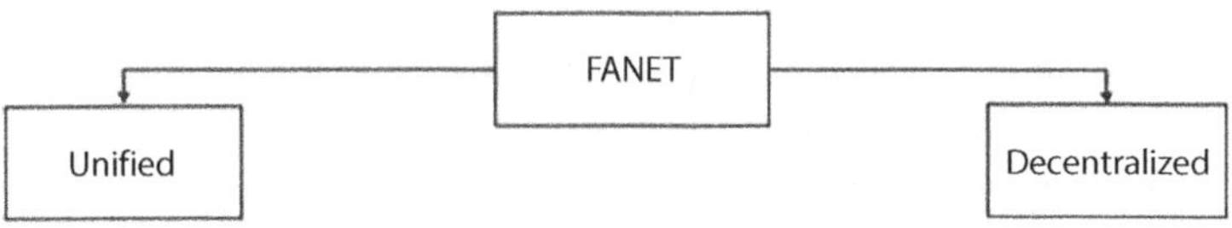

Figure 6.3 Types of FANETs.

6.2.1.1 Challenges in FANET Activity

The principal challenges in the activity of FANETs are:

a) **Restricted data transmission:** The transmission capacity accessible between UAVs is restricted, particularly in the air-to-air (A2A) joins. This can restrict how much information that can be moved between UAVs.
b) **Mobility:** Because they are mobile, UAVs' links to other UAVs can frequently break. This can make it hard to keep a steady organization association.
c) **Interference:** Radio signals between UAVs can be disrupted. The A2A links' range and dependability may suffer as a result.

Regardless of these difficulties, FANETs are a promising innovation for various applications. They are already being used for military operations, disaster relief, and surveillance. FANETs are likely to be used for even more applications in the future as the technology advances.

6.2.1.2 Activity of FANETs

Here are a few extra insights regarding the activity of FANETs:

a) **Protocols for routing:** The *ad hoc* on-demand distance vector protocol is typically the foundation for the routing protocols that are utilized in FANETs. It is a dynamic steering convention that is intended for impromptu organizations. It can adjust to the changing geography of a FANET as UAVs move around [5].
b) **Security:** Security is a central issue in FANETs. The UAVs might be conveying delicate information, and they might be defenseless against assault from malevolent entertainers. Data is protected in FANETs by security measures like encryption and authentication.
c) **Energy conservation:** In FANETs, energy efficiency is another important factor to think about. Because UAVs only have a limited amount of battery power, it is critical that they use as little energy as possible for communication. The FANETs are made more energy efficient by using data compression and routing protocols that save energy.

6.2.2 Explanation of FANET Architecture and Operation

6.2.2.1 Key Characteristics and Difficulties of FANETs

6.2.2.1.1 Key Characteristics shown by Figure 6.4 as follows:

a) **High portability:** UAVs are versatile, and that implies that their connections to other UAVs can break as often as possible. This can make it challenging to keep a steady organization association.

b) **A lack of bandwidth:** The transfer speed accessible between UAVs is restricted, particularly in their A2A joins. This can restrict how much information can be moved between UAVs.

c) **Dynamic organization geography:** A FANET's topology can change quickly as UAVs move around. This can make it challenging to course information bundles between UAVs.

d) **Natural impacts:** The climate can fundamentally affect the exhibition of a FANET. Factors like climate, territory, and electromagnetic impedance can all corrupt the presentation of the organization.

6.2.2.1.2 Challenges

a) **Routing:** In FANETs, routing is a significant obstacle. Finding effective routes for data packets is difficult due to the limited bandwidth and dynamic topology of the network.

b) **Security:** Security is a main pressing issue in FANETs. The UAVs might be conveying delicate information, and they might be helpless against assault from vindictive entertainers.

c) **Energy effectiveness:** Energy productivity is one more significant thought in FANETs. Unmanned aerial vehicles have restricted battery power, so it is vital to limit how much energy they use for correspondence.

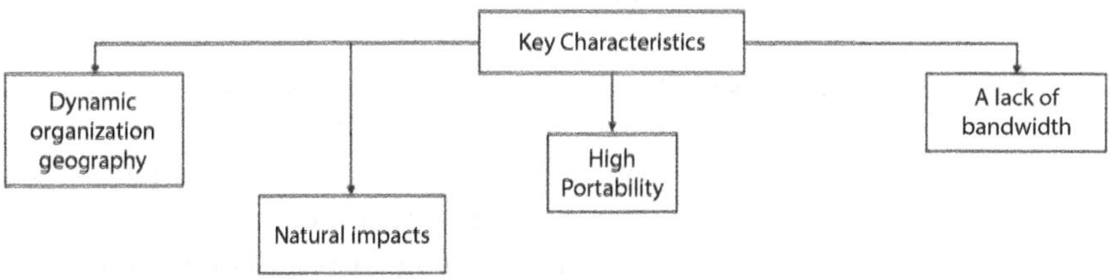

Figure 6.4 Key characteristics of a FANET.

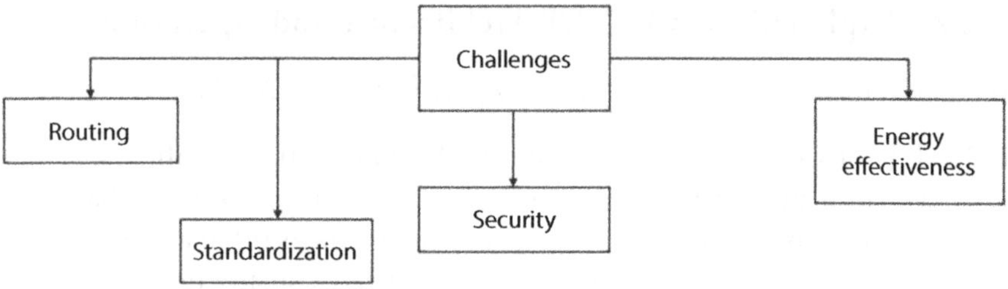

Figure 6.5 Challenges in a FANET.

d) **Standardization:** There is no single norm for FANETs. Interoperating between different FANETs may be challenging as a result of this.

Figure 6.5 show different challenges in FANETs which are a promising technology for a wide range of uses despite these obstacles. They are already being used for military operations, disaster relief, and surveillance. As the innovation keeps on creating, FANETs are may be utilized for more applications later.

6.2.3 Use Cases and Advantages of a FANET in Various Industries

6.2.3.1 Applications of a FANET

The FANET can be used in a lot of different fields. The following are some significant applications and benefits of a FANET:

a) **Protection and surveillance:** A swarm of drones can carry out coordinated security and surveillance tasks thanks to a FANET. Drones furnished with cameras and sensors can cover huge regions, share constant video feeds, and trade data about possible dangers. The FANET considers effective observations of foundations, borders, catastrophe zones, and public occasions.
b) **Rescue and search:** In search-and-rescue operations, FANETs can be great assistance. Drones with thermal, infrared, and high-resolution cameras can quickly scan a large area, find people who are missing or stranded, and tell rescue teams where they are. The ongoing correspondence and

coordination capacities of FANET upgrades the proficiency and viability of search-and-salvage missions [6].

c) **Catastrophe zones:** In scenarios involving disaster management, a FANET is very important. Drones outfitted with correspondence and detecting capacities can assemble information about impacted regions, evaluate harms, and hand off basic data to crisis reaction groups. Drone deployment and coordination are made possible by FANETs, making it easier to efficiently assess disasters, allocate resources, and make decisions.

d) **Accuracy farming:** a FANET can alter the agrarian business by empowering accuracy farming procedures. Data on crop health, pest infestations, and soil moisture can be gathered by drones with a variety of sensors. By sharing this information through a FANET, ranchers can settle on informed choices with respect to the water system, treatment, and bug control, prompting advanced crop yields and diminished asset utilization.

e) **Delivery of packages:** A FANET offers significant advantages in the delivery of packages. Drones outfitted with freight compartments can shape an organization to move bundles effectively, starting from one area to the next. A FANET works with course advancement, snag aversion, and ongoing correspondence among conveyance drones, guaranteeing ideal and secure bundle transportation.

6.2.3.2 *Benefits of a FANET*

a) **Adaptability and versatility:** Drones can dynamically join or leave the network without being disrupted thanks to a FANET's extreme adaptability and scalability. This adaptability empowers a proficient and coordinated effort among an enormous number of robots, adjusting to changing mission prerequisites and organization conditions.

b) **Overt repetitiveness and vigor:** A FANET's cross-section network geography guarantees overt repetitiveness and heartiness. The network is able to automatically reroute data through alternative routes in the event of a drone or communication link failure, ensuring reliable connectivity and data transmission [7].

c) **Operational independence:** Drones are given the ability to make decisions on their own thanks to a FANET. Without relying on centralized control, each drone can independently make decisions about where to go, adjust to changing conditions in the network, and work with other drones. A decentralized and efficient operation is made possible by this autonomy.

d) **Fast sending:** Drones can be quickly deployed in a variety of situations thanks to a FANET. Drones do not require any pre-existing infrastructure to set up a network. This speedy arrangement capacity makes a FANET appropriate for time-basic missions, like crisis reaction or reconnaissance activities.

e) **Cost-viability:** A FANET can be a financially savvy arrangement contrasted with customary foundation subordinate organizations. It eliminates the need for extensive ground-based infrastructure like base stations or communication towers, lowering deployment costs and making it possible to operate in inaccessible or remote locations.

The FANET has revolutionized the capabilities of UAVs and made it possible for advanced communication and collaboration in the air thanks to its inherent advantages and wide range of applications.

6.3 Introduction to the IoT

The IoT is an arrangement of interrelated registering gadgets, mechanical and computerized machines, items, creatures or individuals that are given special identifiers and the capacity to move information over an organization without expecting human-to-human or human-to-PC cooperation.

The IoT is a broad term that includes smart homes, wearable devices, industrial automation, and smart cities.

The ongoing idea between all IoT applications is the utilization of sensors and actuators to gather and communicate information over an organization.

The IoT is still in its beginning phase of improvement; however, it is developing quickly. The quantity of associated IoT gadgets is supposed to arrive at 50 billion by 2025. This development is driven by various variables, including the diminishing expense of sensors and actuators, the rising accessibility of remote network, and the developing interest for information driven experiences.

The IoT has the potential to change many industries. For instance, it may be used to improve the effectiveness of assembling, to computerize transportation, and to give better medical care administrations. The IoT can likewise be utilized to make new items and administrations that were impractical previously.

Be that as it may, the IoT likewise raises various difficulties, like security, protection, and the administration of information. To guarantee that the IoT can be used safely and responsibly, these obstacles must be resolved. The Figure 6.6 show the various Advantages and Disadvantages of the IOT.

6.3.1 Advantages of the IoT

Here are a portion of the advantages of the IoT:

a) **Increased effectiveness:** The IoT can assist in the proficiency of many cycles, like assembling, transportation, and medical care. Sensors, for instance, can be utilized to anticipate potential issues and monitor the performance of equipment. This can help decrease free time and further develop efficiency.

b) **Heightened security:** The IoT can assist with further developing people's wellbeing. For instance, sensors can be utilized to screen the climate for dangers, like breaks or flames. Accidents and injuries can be avoided with this assistance.

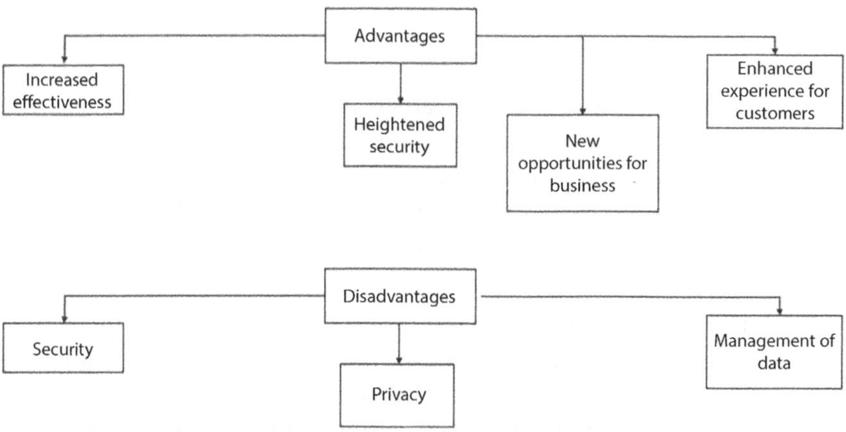

Figure 6.6 Advantages and disadvantages of the IoT.

c) **Enhanced experience for customers:** The IoT can be utilized to upgrade the client experience. For instance, sensors can be utilized to provide customers with real-time updates and track the location of products. This can assist with further developing consumer loyalty.
d) **New opportunities for business:** The IoT has the potential to open new markets for businesses. The IoT can be used to create new goods and services like smart homes and wearable devices, for instance [8].

6.3.2 Difficulties of the IoT

Here are a portion of the difficulties of IoT:

a) **Heightened security:** The IoT is a significant objective for cyberattacks. Sensors and actuators can be hacked to take information or to upset activities. It is vital to execute safety efforts to safeguard IoT gadgets and organizations.
b) **Privacy:** The IoT raises privacy concerns. The information gathered by IoT gadgets can be utilized to follow individuals' developments and exercises. When using IoT devices, it is critical to safeguard individual privacy.
c) **Management of data:** The IoT generates a lot of data. A method for gathering, storing, and analyzing this data is essential. This can be difficult, particularly for smaller businesses.

6.3.2.1 Definition and Core Principles of the IoT

The IoT alludes to the organization of gadgets, vehicles, apparatuses, and different articles installed with sensors, programming, and network capacities that empower them to gather and trade information over the web. These interconnected gadgets can speak with one another, and with incorporated frameworks, make it conceivable to accumulate information, dissect it, and start activities given the knowledge acquired from the information. Figure 6.7 shown the various key principles of IoT.

6.3.2.1.1 Principles of the IoT

Connectivity: The ability of devices to connect to the internet and to one another is one of the fundamental IoT principles. Users can share data and work together on a variety of projects, resulting in a network of connected devices.

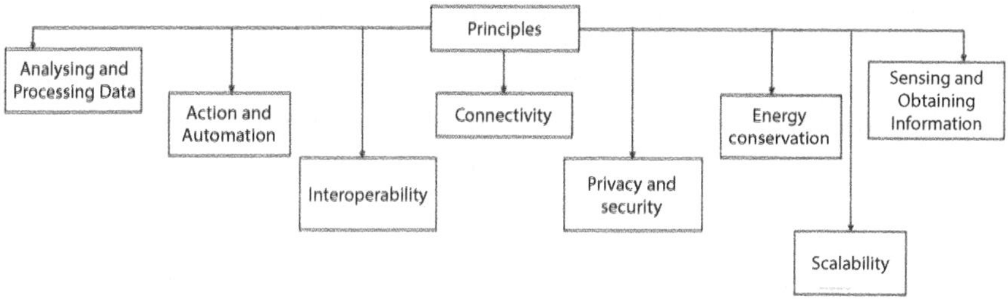

Figure 6.7 Principles of the IoT.

Sensing and obtaining information: IoT devices have a variety of sensors that let them collect information from the environment. Depending on the application and use case, these sensors can measure temperature, humidity, motion, light, pressure, and more [9].

Analyzing and processing data: The IoT entails processing and evaluating the data gathered from connected devices. This might happen locally on the gadget or be communicated to a unified framework for investigation. The investigation can prompt significant bits of knowledge, empowering educated and independent robotization processes.

Action and automation: IoT frameworks frequently expect to mechanize assignments and cycles in light of information. For instance, brilliant home gadgets can change the temperature founded on inhabitance and climate information. Manufacturing processes can be optimized in real-time by industrial IoT applications, reducing waste and increasing productivity.

Interoperability: The IoT must be able to seamlessly integrate systems and devices from various manufacturers to be effective. Interoperability guarantees that gadgets can convey and share information normally, no matter their image or innovation.

Privacy and security: With the rising number of associated gadgets, guaranteeing the security and protection of information becomes basic. The IoT frameworks should execute hearty and safe efforts to safeguard against unapproved access, information breaks, and digital assaults.

Scalability: IoT arrangements frequently include countless associated gadgets, and they ought to be intended to deal with versatility as the number of gadgets and information volume increments over the long haul.

Energy conservation: Numerous IoT gadgets work on battery power or have restricted energy assets. Therefore, energy efficiency is essential for extending the devices' lifespan and reducing the need for upkeep.

6.3.2.2 *Components and Layers of the IoT Ecosystem*

The IoT ecosystem consists of various components and layers that work together to enable the functioning of the IoT. Figure 6.8 is the evidences of various Components and layers of the IoT Ecosystem. Here are the key components and layers of the IoT ecosystem:

Devices/things: Devices or things are the physical objects that are connected to the internet and equipped with sensors, actuators, and communication capabilities. These can include a wide range of objects such as sensors, wearables, appliances, vehicles, industrial machinery, and more. Devices collect data from the environment, interact with users, and execute commands.

Connectivity: Connectivity refers to the means by which IoT devices connect to the internet and communicate with each other. This can be achieved through various technologies such as Wi-Fi, Bluetooth, cellular networks, satellite communication, Zigbee, and LoRaWAN. The choice of connectivity technology depends on factors such as range, power consumption, data transfer rate, and cost.

Data management: Data management involves the collection, storage, processing, and analysis of the massive amounts of data generated by IoT devices. This layer includes technologies and platforms for data storage, data integration, data preprocessing, and data analytics. It encompasses data management techniques like edge computing, cloud computing, and distributed computing to handle the data efficiently.

Cloud platform: The cloud platform is a crucial layer in the IoT ecosystem. It provides scalable computing and storage resources to process and store data from IoT devices. Cloud platforms offer services like data storage, data analytics, machine learning, and application development tools. They enable data aggregation, real-time processing, and advanced analytics capabilities for IoT solutions [10].

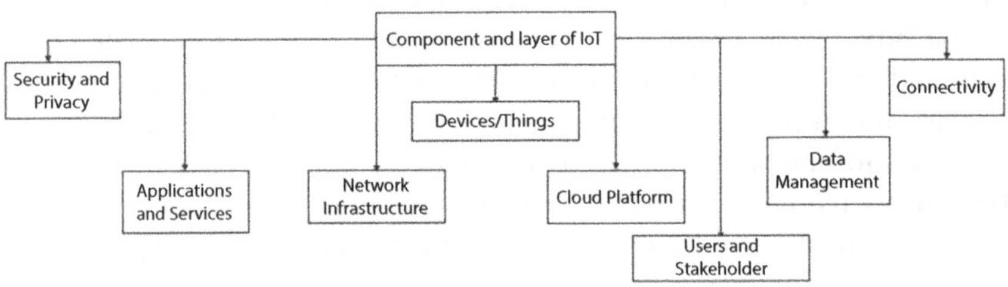

Figure 6.8 Components and layers of the IoT.

Network infrastructure: The network infrastructure layer consists of the underlying network infrastructure that facilitates communication between devices and connects them to the internet. It includes routers, switches, gateways, access points, and other network components that enable data transmission and connectivity. Network infrastructure can be a combination of wired and wireless technologies, including the Ethernet, Wi-Fi, and cellular networks.
Security and privacy: Security is a critical layer in the IoT ecosystem to protect devices, data, and communication from unauthorized access, tampering, and cyber threats. It involves implementing encryption, authentication mechanisms, access controls, secure protocols, and monitoring systems. Privacy considerations involve ensuring that user data is handled securely and transparently, with consent and compliance with relevant regulations.
Applications and services: The top layer of the IoT ecosystem includes applications and services that utilize the data generated by IoT devices. These applications can range from simple mobile apps to complex enterprise solutions. Examples include smart home automation, industrial monitoring, asset tracking, healthcare monitoring, and environmental sensing. Applications and services leverage the data insights to provide value-added functionalities and enable automation and decision-making [11].
Users and stakeholders: The IoT ecosystem involves users who interact with IoT devices and applications. Users can be individuals, businesses, or organizations. Stakeholders include manufacturers, developers, service providers, and regulatory bodies that play a role in the design, deployment, and governance of IoT solutions.

6.4 Internet-of-Things-Enabled Communication in a FANET

The IoT-enabled communication in a FANET refers to the integration of IoT devices within the network of UAVs or drones. It allows for the exchange of data and information between IoT devices on the ground and the drones in the sky. This integration can provide additional capabilities and enhance the functionality of FANET.

Sensor integration: IoT devices on the ground can be equipped with sensors that collect data such as environmental conditions, air quality, temperature, humidity, or other relevant parameters. These IoT sensors can transmit real-time data to the drones in the FANET network, providing them with valuable information for decision-making and mission planning.

Data exchange: IoT-enabled communication allows for the exchange of data between the drones and ground-based IoT devices. Drones can transmit data such as aerial images, video feeds, or telemetry data to the IoT devices. Similarly, ground-based IoT devices can transmit data collected by their sensors to the drones. This bi-directional data exchange enhances situational awareness, monitoring capabilities, and collaborative operations.

Remote control and command: IoT-enabled communication can facilitate remote control and command of drones within the FANET network. Ground-based IoT devices can send commands and instructions to the drones, enabling remote operation, waypoint setting, or task assignment. This allows for centralized control and coordination of drone operations from the ground.

Data aggregation and analytics: IoT-enabled communication in FANET enables the aggregation of data collected by drones and ground-based IoT devices. This data can be transmitted to a centralized system or cloud platform for further processing, analysis, and visualization. Advanced analytics techniques can be applied to the aggregated data to gain insights, detect patterns, and make informed decisions.

Intelligent automation: By integrating IoT devices within a FANET, intelligent automation capabilities can be enhanced. Drones can receive commands and instructions from IoT devices, enabling them to autonomously perform specific tasks or respond to changing conditions. For example, drones can adjust their flight paths based on real-time data received from IoT sensors on the ground [12].

Enhanced mission capabilities: IoT-enabled communication Figure 6.9 expands the mission capabilities of a FANET. By integrating with IoT devices, drones can leverage the additional sensors, data sources, and processing capabilities available on the ground. This enables them to perform more complex missions, such as environmental monitoring, precision agriculture, or infrastructure inspection, by leveraging the IoT ecosystem.

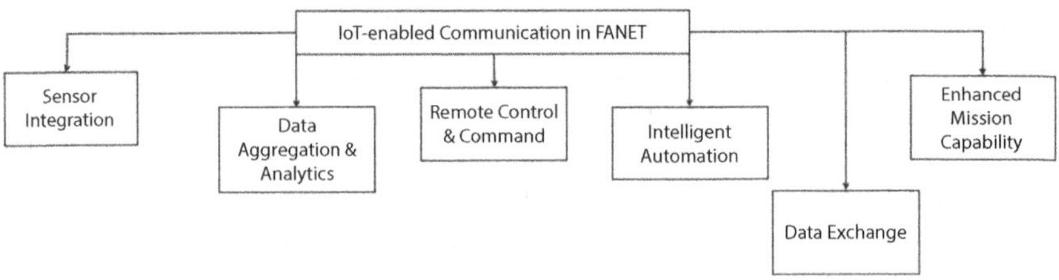

Figure 6.9 Internet-of-things-enabled communication in a FANET.

Overall, IoT-enabled communication in FANET enhances the capabilities of both the drones and ground-based IoT devices by enabling seamless data exchange, remote control, and intelligent automation. This integration opens new possibilities for collaborative operations, data-driven decision-making, and advanced mission capabilities in various domains [13].

6.5 Internet-of-Things Communication Protocol

The IoT communication protocol is suitable for FANETs (e.g., MQTT, CoAP). A FANET refers to a network of UAVs that communicate with each other and potentially with ground-based stations to enable coordinated and autonomous flight operations. When it comes to IoT communication protocols suitable for FANET, there are several options to consider. Two commonly used protocols are the message queuing telemetry transport (MQTT) and constrained application protocol (CoAP). Let us explore these protocols in more detail:

6.5.1 Message Queuing Telemetry Transport

- The MQTT is a lightweight and publish-subscribe-based messaging protocol that is widely used in IoT applications.
- It operates on top of the TCP/IP protocol, making it reliable and suitable for scenarios where network connections are relatively stable.
- The MQTT is known for its low overhead, making it suitable for resource-constrained devices and networks with limited bandwidth.
- It supports a flexible message structure, allowing devices to publish and subscribe to topics, enabling efficient data exchange between devices.
- The MQTT also supports quality-of-service levels to ensure message reliability and delivery guarantees.

6.5.2 Constrained Application Protocol

- The CoAP is a specialized web transfer protocol designed for resource-constrained devices and networks, making it well-suited for IoT applications.

- It operates on top of the user datagram protocol (UDP), which is a lightweight transport protocol that provides reduced overhead compared to TCP/IP.
- The CoAP uses a request-response model, like the HTTP, allowing devices to retrieve and manipulate resources through simple methods such as GET, PUT, POST, and DELETE.
- It supports multicast communication, which is beneficial for FANET scenarios where multiple UAVs need to receive the same information simultaneously.
- The CoAP also provides support for resource discovery, making it easier for devices to locate and interact with each other within the FANET.

Both the MQTT and CoAP have their strengths and can be suitable for FANET applications, depending on the specific requirements and constraints of the network. The MQTT is often preferred for scenarios with more stable network connections and when reliability and message persistence are critical. On the other hand, the CoAP is a good choice for resource-constrained environments, low-power devices, and situations where multicast communication and resource discovery are important [14].

6.5.3 Data Aggregation and Routing Strategies in IoT-Enabled FANET

Information total and directing methodologies are significant contemplations for IoT-empowered FANETs. The information total is the method involved with gathering information from different sources and summing it up into a more modest and sensible structure. Steering is the method involved with deciding the best way for information to head out from its source to its objective [15, 16].

6.5.3.1 Methodologies of IoT-Enabled FANETs

In IoT-enabled FANETs, a variety of data aggregation and routing strategies are available. The absolute most normal methodologies include:

A centralized collection of data: A central node collects all data and then aggregates it in this manner. This is the simplest and most effective strategy, but if the central node becomes overloaded, it could become a bottleneck.

Accumulation of distributed data: In this procedure, information is collected by numerous hubs in the organization. This can further develop versatility and unwavering quality, yet it can likewise be more intricate to carry out.
Various leveled information: Data is gathered in a hierarchical fashion using this strategy. This can further develop adaptability and proficiency; however, it can likewise be more intricate to execute. The selection of information conglomeration procedure relies upon the particular application and the necessities of the organization [17].

6.5.3.2 *Methodologies of IoT-Empowered FANETs*

There are different steering techniques that can be utilized in IoT-empowered FANETs. The absolute most normal methodologies include:

The shortest route: The shortest route between the source and destination nodes is the goal of this strategy. Although this is the most effective tactic, putting it into action in a dynamic network like a FANET can be challenging.
Routing hop by hop: In this methodology, information is directed from one hub to another until it arrives at its objective. This is a straightforward procedure that is not difficult to carry out, yet it may not be the most productive.
Proactive steering: In this methodology, hubs keep up with directing tables that store data about the best ways to different hubs in the organization. While this may increase productivity, its implementation may be more difficult.

The specific application and the network's requirements determine the routing strategy to use.

IoT-enabled FANETs must take data aggregation and routing into account. By cautiously picking the right procedures, working on the presentation and adaptability of the network is conceivable [18].

6.6 Conclusion

6.6.1 Recap of the Key Points

Explicit substances or sections are not approached by the AI language model. Notwithstanding, here is an outline of the central issues that may be shrouded in a part on the coordination of the IoT in FANETs:

a. **Prologue to a FANET:** Give an outline of FANET, which is a decentralized organization of independent robots or automated flying vehicles (UAVs) that speak with one another to frame an impromptu organization.
b. **Internet-of-Things presentation:** Momentarily make sense of what the IoT is and its importance in associating ordinary items and gadgets to the web, empowering information trade and robotization [19].
c. **Challenges in a FANET:** Examine the interesting difficulties looked at by a FANET, like its versatility, restricted energy, restricted data transmission, and the requirement for effective correspondence conventions.
d. **Coordination of the IoT in a FANET:** Make sense of how IoT innovation can be coordinated into a FANET to improve its abilities and address a portion of the difficulties. This could incorporate furnishing drones with IoT sensors, actuators, and correspondence modules.
e. **Internet-of-Things-empowered applications in a FANET:** Investigate different applications where IoT joining can be helpful in a FANET. This might incorporate reconnaissance, ecological checking, horticulture, and conveyance administrations.
f. **Correspondence conventions:** Examine the correspondence conventions utilized in a FANET for IoT gadgets to cooperate with one another and with ground-based IoT foundation.
g. **Information, the executives, and investigation:** Make sense of how information produced by IoT gadgets in FANET can be gathered, handled, and dissected to extricate important bits of knowledge and backing navigation [20].
h. **Security and protection:** Address the security and protection concerns connected with coordinating IoT in FANET, as robots gather and send touchy information.
i. **Energy proficiency:** Feature the significance of energy productivity in FANET, and how IoT joining can advance robot tasks to monitor energy.
j. **Future patterns:** Talk about possible future improvements in the joining of IoT in FANET, remembering headways for innovation and new use cases.

6.6.2 Potential Impact of IoT Integration on the Future of FANET

The integration of the IoT in FANETs holds immense promises for shaping the future of autonomous drone technology. By equipping drones with IoT sensors and communication modules, a multitude of possibilities emerge, driving the evolution of FANET to new heights. One of the most significant impacts is the ability to enhance data collection and analysis. Internet-of-Things-enabled drones can gather real-time information on environmental conditions, weather patterns, and other critical data, enabling better decision-making and situational awareness. This enhanced coordination and autonomy among drones facilitate seamless cooperation on complex tasks such as search-and-rescue operations, environmental monitoring, and surveillance. Additionally, integration empowers drones to monitor their own health, battery levels, and performance, optimizing resource management, and prolonging operational life. Moreover, IoT integration paves the way for a wider range of applications, spanning precision agriculture, urban planning, infrastructure inspection, and even smart city delivery services. Furthermore, the FANET's IoT-enabled drones can extend communication networks to remote and disaster-stricken areas, bolstering emergency response capabilities and supporting connectivity needs. However, these advancements also come with challenges, including data management, security, privacy, and the need for updated regulations. As the technology matures, addressing these challenges will be crucial to unlock the full potential of IoT integration in a FANET, ushering in a new era of aerial connectivity and transformative applications across various industries and domains.

Bibliography

1. Kühn, W. and Grell, M., *JDF, Process Integration, Technology, Product Description*, Springer, 2005. ISBN 3-540-23560-4.
2. Toumi, K., Ayari, M., Saïdane, L.A., Bouet, M., Pujolle, G., HAT: HIP Address Translation Protocol for Hybrid RFID/IP Internet of Things Communication. *2010 International Conference on Wireless and Ubiquitous*, 2010.
3. Song, M., Xu, Y., Zhang, B., Cao, S., Research and Design of the MicroBlog Content Processing System Based on Integration of Internet of Things and Social Media. *Appl. Mech. Mater.*, 411–414, 317–321, 2013. 10.4028/www.scientific.net/AMM.411-414.317.

4. Datta, S.K. and Bonnet, C., Describing Things in The Internet of Things: From CoRE Link Format to Semantic Based Descriptions. *2016 IEEE International Conference and Consumer*, 2016, (IF: 3).
5. Kokoris-Kogias, E., Voutyras, O., Varvarigou, T.A., TRM-SIoT: A Scalable Hybrid Trust & Reputation Model for The Social Internet of Things. *2016 IEEE 21st International Conference on Emerging*, 2016, (IF: 3).
6. Botta, A., Donato, W., Persico, V., Pescapè, A., Integration of Cloud computing and Internet of Things: A survey. *Future Gener. Comput. Syst.*, 56, 2015. 10.1016/j.future.2015.09.021.
7. Sidid, S. and Gaur, S., Smart Grid Building Automation Based on Internet of Things. *2017 Innovations in Power and Advanced Computing*, 2017.
8. Yu, J., Kwon, SH., Kang, H., Kim, S.-J., Bae, J.-H., Pyo, C.-S., A Framework on Semantic Thing Retrieval Method in IoT and IoE Environment. *2018 International Conference on Platform Technology and*, 2018.
9. Acharya, A., Schroeder, D., Schwei, K., Chyou, P.H., Update on electronic dental record and clinical computing adoption among dental practices in the United States. *Clin. Med. Res.*, 15, 3–4, 59–74, 2017. doi: 10.3121/cmr.2017.1380. [DOI] [PMC free article] [PubMed] [Google Scholar]
10. Jadala, V.C., Pasupuletti, S.K., Raju, S., Kavitha, S., Bhaba, CM.H.S., Sreedhar, B., Need of Intenet of Things, Industrial IoT, Industry 4.0 and Integration of Cloud for Industrial Revolution. *2021 Innovations in Power and Advanced Computing*, 2021.
11. Uckelmann, D., Harrison, M., Michahelles, F., An Architectural Approach Towards the Future Internet of Things, 2011. 10.1007/978-3-642-19157-2_1.
12. Colitti, W., Steenhaut, K., De Caro, N., Buta, B., Dobrota, V., REST Enabled Wireless Sensor Networks for Seamless Integration with Web Applications. *2011 IEEE Eighth International Conference on Mobile Ad-Hoc*, 2011, (IF: 4).
13. Mehaseb, M.A., Gadallah, Y., El-Hennawy, H.M., WSN Application Traffic Characterization for Integration Within The Internet of Things. *2013 IEEE 9th International Conference on Mobile Ad-Hoc ands*, 2013, (IF: 3).
14. Vilarinho, T., Farshchian, B.A., Floch, J., Mathisen, B.M., A Communication Framework for The Internet of People and Things Based on The Concept of Activity Feeds in Social Computing. *2013 9th International Conference on Intelligent*, 2013, (IF: 3).
15. Jayavel, K. and Nagarajan, V., Survey of Migration, Integration and Interconnection Techniques of Data Centric Networks to Internet- Towards Internet of Things (IoT). *Ind. J. Sci. Technol.*, 9, 2016. 10.17485/ijst/2016/v9i8/87924.
16. Stergiou, C. Psannis, K. Kim, B.-G., Gupta, B.B., Secure integration of IoT and Cloud Computing. *Future Gen. Comput. Syst.*, 78, 2016.
17. Molano, J.I.R., Lovelle, J.M.C., Montenegro, C.E., Granados, J.J.R., Crespo, R.G., Metamodel for integration of Internet of Things, Social Networks, the Cloud and Industry 4.0. *J. Ambient Intell. Humaniz. Comput.*, 9, 3, 709–723, 2017. https://doi.org/10.1007/s12652-017-0469-5

18. Barros, V., An Approach for the Integration of Different Data Sources from Sensor Networks in the Internet of Things Context, 2019. 10.13140/RG.2.2.23484.59524.
19. Chen, M., Miao, Y., Humar, I., *OPNET IoT Simulation*, Springer Nature, Singapore, 600 pages, April 2019.
20. Voumick, D., Deb, P., Khan, M., Operation and Control of Microgrids Using IoT (Internet of Things). *J. Softw. Eng. Appl.*, 14, 418–441, 2021. 10.4236/jsea.2021.148025.

7

Enhancing Precision Agriculture Through Bio-Inspired Routing Protocols for Flying *Ad Hoc* Networks

S. Nandhini* **and K. S. Jeen Marseline**

Department of IT & Cognitive Systems, Sri Krishna Arts and Science College, Coimbatore, India

Abstract

Integrating modern technology into agriculture is crucial to address the demands of a growing global population. Precision agriculture is a promising solution, which focuses on optimizing crop cultivation and reducing environmental effects. However, this approach faces challenges regarding connectivity and data acquisition in remote and difficult-to-access agricultural regions. A potential solution to these challenges lies in integrating flying *ad hoc* networks (FANETs) into precision agriculture. A FANET consists of unmanned aerial vehicles (UAVs) working to establish communication networks. This integration holds promise for revolutionizing precision agriculture through real-time data collection, monitoring, and rapid deployment across various agricultural terrains. The real-time implementation of a FANET in agriculture introduces complex communication routing issues. The dynamic movement of UAVs, coupled with ever-changing network structures and energy constraints, necessitates robust and adaptive communication protocols. This paper presents an analysis of the various routing protocols used for efficient communication among the UAVs. It emphasizes the application of bio-inspired routing protocols for a FANET.

***Keywords*:** Bio-inspired routing, FANET, precision agriculture, UAVs

**Corresponding author*: nandhinis@skasc.ac.in

Taskeen Zaidi, Adarsh Kumar, Saurav Mallik and Keshav Kaushik (eds.) Autonomous Flying Ad-Hoc Networks, (149–164) © 2025 Scrivener Publishing LLC

7.1 Introduction

Precision agriculture (PA) is a technique that focuses on farm management with the aid of information technology. Different types of technologies are used in field monitoring like GPS sensors, drones, and robots. Precision agriculture aims to find out the variability in agricultural fields. Based on this, suitable crops can be grown to increase productivity. Finding field variability also helps with the optimal use of fertilizer and other inputs for the crops. To monitor the growth of crops, sensors and high-resolution cameras are mounted on unmanned aerial vehicles (UAVs), which are commonly known as drones. The images captured from the UAVs are analyzed to monitor the growth of crops. Based on this, various agricultural inputs like water, fertilizer, and pesticides can be used economically, depending on the nature of the site [1].

When multiple flying UAVs are communicating and exchanging messages, the network is called a flying *ad hoc* network (FANET). There are different types of drones used for farming. They are fixed-wing drones, mullti-rotatary drones, and vertical or landing drones. Fixed-wing drones are commonly used in crop monitoring and surveying [2]. Since they can fly for a longer duration, they cover large areas efficiently even in adverse weather conditions. These types of UAVs are used to analyze the health of crops by taking aerial images of the land with high resolution. The growth of crops can be monitored and based on the variation of nutrient level of soil. Irrigation and other soil management tasks can be done. It also helps to identify the diseases and pests that may affect the growth of crops. Single-rotor drones are another type of UAV, used for various tasks like soil analysis, irrigation management, crop monitoring, and pest management. Compared to fixed-wing drones, single-rotor drones can perform precise operations but are expensive and complex in their operation. Multi-rotor drones are the next category of drones used for close-range crop operations. They are capable of flying in any direction, collecting precise data and producing high-resolution images. Despite these features, it is economical [3]. Hybrid drones are another type of UAV that combines the features of fixed-wing and multi-rotor types. These types of drones have vertical take-off and landing capabilities even in terrain and constrained spaces. They can also operate in any type of weather. Hence, they are more suitable for large-scale operations.

Routing protocols are essential for the effective communication and delivery of messages between nodes. Developing routing protocols for FANETs is a challenging task due to several features of FANETs like high mobility, limited resources, and frequent changes in the topology [4]. Hence, in-depth knowledge about the routing protocols for FANETs is required.

This chapter analyzes bio-inspired routing protocols of FANETs for effective communication. The rest of the paper is organized as follows: An overview of PA, its significane, current technological advancements, and challenges in PA is discussed in Section 7.2. Various bio-inspired routing protocols and the role of FANETs in data collection for PA is analyzed in Section 7.3. Some case studies and real-world examples of FANETs using bio-inspired routing protocols for PA is provided in Section 7.4. The conclusion of the study is presented in Section 7.5.

7.2 Precision Agriculture

Precision agriculture is a farm management strategy that uses different techniques to collect data about crops and fields. This data is then used to make decisions about how to manage the farm more effectively. Precision agriculture aims to increase efficiency, productivity and profitability while reducing environmental impact. Precision agriculture uses several key technologies. The GPS and mapping systems are used to create maps of fields, which can then be used to identify areas of variability. Sensors collect data about a variety of factors, such as soil moisture, nutrient levels, and crop health. Remote sensing technologies, such as satellite imagery and aerial photography, can be used to collect data about crops and fields from a distance [5]. Variable rate application equipment is used to apply inputs like fertilizer, pesticides, and irrigation water, at different rates in different parts of a field. Autonomous vehicles are used in some PA applications for scouting fields [6].

Precision agriculture gives various benefits to farmers, such as increased crop yields, improved crop quality, reduced input costs, and reduced environmental impact. By addressing specific needs of different field zones, PA fosters optimal crop growth conditions, leading to higher yields [7]. Targeted nutrient and pest management practices result in healthier crops with improved quality. Data-driven decision-making minimizes the waste of resources like fertilizers and pesticides, leading to cost savings. Precision agriculture promotes practices that reduce water usage, minimize chemical run-off, and improve soil health, contributing to a more sustainable agricultural ecosystem.

There are technological advancements in PA. Remote sensing technologies like satellites, drones, and multispectral imaging are leading the charge.

They provide precise data on crop health, nutrient deficiencies, and potential yield variations across vast areas. Sophisticated sensors are becoming increasingly affordable and diverse [8]. They continuously monitor soil moisture, nutrient content, temperature, and even plant stress indicators, providing real-time insights into crop health. The massive amount of data collected necessitates advanced software and analytics platforms. Here, artificial intelligence (AI) is used as an enabing technology. Artificial intelligence algorithms can analyze vast datasets, identify complex patterns, and make data-driven recommendations for resource management, optimizing fertilizer application, irrigation, and pest control. Autonomous vehicles and robots are finding their way into PA applications. Imagine drones scouting fields for pests or robots precisely applying herbicides. These innovations promise to enhance efficiency and reduce labor costs [9].

Even though there are many techhnological advancements, some challenges also need to be addressed here. While technology is advancing, the initial investment in high-tech equipment, software, and training can be significant, especially for small-scale farmers. Making PA more cost-effective is crucial for a wider adoption [10]. The sheer volume of data generated by PA systems can be overwhelming. Farmers need user-friendly interfaces and training to manage and utilize this data effectively. Additionally, data security and privacy concerns need to be addressed. The lack of uniform standards for data collection and communication between different PA technologies can create compatibility issues. Establishing common standards will ensure seamless integration and data sharing across platforms. However, in rural areas poor internet connectivity can affect the adoption of cloud-based PA solutions. Bridging the digital divide is essential to ensure equitable access to this technology [11].

Despite the challenges, PA holds immense potential for the future of farming. By continuously developing cost-effective technologies, addressing data management concerns, and promoting digital inclusion, PA can empower farmers to optimize yields, minimize environmental impact, and ensure a sustainable food supply for a growing population [12, 13].

7.3 Bio-Inspired Routing Protocols for FANET

Flying *ad hoc* networks are mobile networks created by drones (UAVs) communicating directly with each other, without relying on a central hub or ground station. These drones can fly autonomously and adjust their positions to maintain connectivity within the network. A network of drones can efficiently gather data across a large field. This data could include high-resolution images for crop health monitoring, multispectral data for nutrient deficiency detection, or real-time soil moisture readings from on-board sensors [14].

Flying *ad hoc* networks can be used to coordinate the movement and spraying of drones for targeted pesticide or fertilizer application. This reduces waste and environmental impact compared to traditional blanket spraying methods. Drones equipped with thermal imaging sensors can fly over fields, identifying areas with varying water needs. This data can be shared within the FANET to optimize irrigation plans and ensure efficient water use [15]. There are several advantages in using FANETs for agriculture. The flying nodes can be increased or decreased depending on the size of the field being monitored. They can navigate uneven terrain and reach remote areas inaccessible to ground vehicles.With FANETs, real-time data collection and sharing is possible. As drone technology continues to evolve, with longer flight times, improved communication protocols, and potentially AI-powered decision making onboard, FANETs hold immense promise for revolutionizing how we manage agricultural land. They offer potential for more efficient, data-driven, and sustainable farming practices, contributing to food security for a growing global population [16, 17].

The defining characteristic of a FANET is the mobility of its nodes. Drones can fly freely, constantly changing their positions and network topology. Flying *ad hoc* networks are self-organizing networks [18]. The drones can establish and maintain connections with each other without needing any external infrastructure. Due to the movement of drones, the network topology of a FANET is constantly changing. Routing protocols need to be adapted to handle these dynamic changes. The communication range between drones is typically limited compared to terrestrial networks. Factors like antenna type and signal strength play a role. Drones have limited battery life, which imposes constraints on network operations. Efficient communication protocols and potential in-air charging solutions are crucial [19–21].

Traditional routing protocols designed for static networks like mobile *ad hoc* networks are not ideal for FANETs due to the unique challenges FANETs present. As the nodes in a FANET move constantly, there are frequent changes in the network topology. The routing protocols designed for other networks struggle to adapt to such rapid changes. Drones typically have a shorter communication range compared to ground-based nodes. This can lead to frequent route breaks and disconnections. Battery life is a major concern for drones. Routing protocols need to be energy-efficient to maximize network operation time. Proactive routing protocols, reactive routing protocols, and hybrid routing protocols are the major categories of routing protocols [22–24]:

- **Proactive routing protocols:** These protocols work by maintaining routing tables with pre-computed paths between nodes. This ensures minimal delay in data delivery, but the constant network changes in FANETs can make these tables outdated quickly, wasting energy on maintaining them. Examples include the open shortest path first.
- **Reactive routing protocols:** These protocols establish routes only when needed, on-demand. This reduces control overhead and conserves energy. However, delays can occur during route discovery, especially in highly dynamic networks. Examples include *ad hoc* on-demand distance vector (AODV) routing.
- **Hybrid routing protocols:** These protocols contain the features of both proactive and reactive approaches. They might maintain routes for frequently accessed destinations while using on-demand route discovery for less frequently accessed ones. This offers a balance between route stability and energy efficiency. Examples include ZRP (Zone Routing Protocol).

These traditional routing protocols often struggle in FANETs due to high node mobility, limited communication range, and energy constraints. Bio-inspired approaches offer potential solutions to these challenges. They can adapt to changing network conditions, mimicking the way natural systems adjust to their environment. They can efficiently handle networks with a varying number of drones, similar to how biological systems operate effectively at different scales [25]. They can remain functional even with disruptions or node failures, inspired by the inherent resilience of biological systems. Bio-inspired protocols can dynamically adjust to network changes, leading to more stable and efficient routing. By mimicking natural communication patterns, these protocols can minimize the number of control messages exchanged, conserving energy. They can efficiently handle networks of varying sizes and remain functional even with disruptions [26].

There are many prominent bio-inspired routing approaches for FANETs. They are based on the social behavior and foraging of animals, insects and microbes [27–29].

7.3.1 Gray Wolf Optimization

It is a fascinating population-based optimization algorithm inspired by the leadership hierarchy and hunting strategies of gray wolves. It is a powerful tool used to solve complex optimization problems in various fields, including agriculture. They live in packs with a strict social hierarchy. An alpha wolf leads the pack, followed by beta, delta, and omega wolves. Wolves rely on teamwork and coordinated strategies to hunt prey. They track, chase, and surround their target before attacking [30]. Gray wolf optimization (GWO) is used to optimize several PA features like crop yield prediction, irrigation scheduling, resource allocation. Optimizing models help to predict the crop yield based on several factors like weather, soil conditions, and fertilizer application. The algorithm helps to determine the optimal irrigation plan to maximize crop yield while minimizing water usage. It also helps in optimizing the allocation of resources like fertilizer and pesticides across different zones within a field. Gray wolf optimization is a powerful tool with exciting applications in PA. By mimicking the social behavior and hunting strategies of gray wolves, GWO is useful for farmers to make decisions based on the data, to optimize resource use and to improve agricultural outcomes [31].

7.3.2 BAT Algorithm

The BAT algorithm is another interesting optimization technique used in PA, inspired by the echolocation abilities of bats. Microphones are essential for humans, but bats navigate and hunt using echolocation [32]. They emit high-frequency sound waves and analyze the echoes bouncing back from objects to perceive their surroundings. The BAT algorithm can be applied to various optimization problems in PA. The placement of sensors in a field helps to collect the most relevant data for crop health monitoring. Optimizing algorithms help for early detection of crop diseases based on sensor data. Optimizing the allocation of resources like fertilizer and water based on real-time data and crop needs is also possible [33].

7.3.3 Sand Cat Swarm Optimization Algorithm

This algorithm takes inspiration from the hunting and foraging behavior of sand cat. They are capable of chasing and hunting their prey, even underground, using their ability to hear noise with frequency below 2 kHz [34]. The searching and hunting behavior of a swarm of sand cats is reflected in the sand cat swarm optimization algorithm. It is a metaheuristic algorithm used to find an optimal solution for the given problem.

7.3.4 Ant Colony Optimization Algorithm

This algorithm is inspired by the foraging behavior of ants. It is a popular tool used to solve complex pathfinding and optimization problems in various fields, and agriculture is a prime example. Ant colony optimization can be used for various optimization tasks in PA as follows [35]:

- **Route planning for field monitoring:** Optimizing the path for drones or ground vehicles to collect data across a field, minimizing travel time and maximizing efficiency.
- **Delivery route optimization:** Planning optimal delivery routes for agricultural supplies or produce, considering factors like distance, load capacity, and road conditions.
- **Resource allocation:** Optimizing the allocation of resources like water or fertilizer across different zones within a field based on specific needs.

7.3.5 Bee Colony Optimization

Also known as the bees algorithm, it is another fascinating optimization technique inspired by the foraging behavior of honey bees. It is a powerful tool used to solve various optimization problems, finding its place in the realm of PA, as well. Honey bees are known for their efficient foraging strategies [36]. Scout bees locate food sources, evaluate their quality (nectar amount), and communicate this information to other bees through a "waggle dance." This dance encodes the distance and direction of the food source, allowing other worker bees to efficiently locate it.

Bee colony optimization (BCO) can effectively find good solutions for complex optimization problems [37]. The employed bees and scout bees promote exploration of the search space, while the onlooker bees focus on exploiting promising areas with high-fitness solutions. The algorithm is easier to understand and implement compared to some other optimization techniques. Bee colony optimization can be applied to various optimization problems in PA, including: crop yield prediction, resource management, planting, and seeding optimization. Here, BCO can help find the combination of factors that maximizes predicted yield. The allocation of resources like water, fertilizer, and pesticides across different zones within a field is done by considering soil type, crop needs, and environmental impact. Determining the optimal planting or seeding patterns within a field to maximize yield or resource utilization is also possible [38–45].

7.3.6 Firefly Optimization Algorithm

It is a bio-inspired optimization technique that draws inspiration from the flashing behavior of fireflies. This algorithm helps to solve complex optimization problems in various fields, and PA is one of them [38–40]. Firefly optimization (FFO) can effectively find good solutions for complex optimization problems. The algorithm is easier to understand and implement compared to some other optimization techniques. Firefly optimization requires minimal parameter tuning compared to some other algorithms [46]. Firefly optimization can help identify the combination of factors that leads to the highest predicted yield. By mimicking the communication and light-based attraction of fireflies, FFO will be helpful for farmers to make decisions for optimal resource use, crop yields, and overall farm management [47].

7.3.7 Case Studies and Real-World Examples of FANET in Precision Agriculture

While large-scale commercial applications of FANETs in PA are still emerging, there are several pilot projects and research initiatives showcasing the potential of this technology.

7.3.7.1 Case Study 1

Researchers at the Chinese Academy of Agricultural Sciences deployed a FANET consisting of multi-rotor drones for monitoring crop health in a large rice field. Drones equipped with multispectral cameras flew preprogrammed routes using a waypoint navigation system (not strictly bio-inspired). They captured high-resolution images used to assess crop health, identify nutrient deficiencies, and detect potential diseases. The FANET enabled rapid and efficient data collection across the vast field, providing valuable insights for targeted interventions and improved crop management.

7.3.7.2 Case Study 2

A research team at the University of Seville investigated using FANETs for precision spraying of vineyards. The project explored using multiple drones equipped with spraying nozzles. These drones would communicate and coordinate their movements using a central control system. This would allow for the targeted application of pesticides or herbicides, minimizing waste and environmental impact. The project highlighted the need for robust communication protocols and real-time obstacle detection to ensure safe and efficient operation within the vineyard.

7.3.7.3 Case Study 3

The Commonwealth Scientific and Industrial Research Organisation in Australia is exploring FANETs for smart irrigation management. The project envisions a network of drones equipped with soil moisture sensors. These drones would fly programmed routes (potentially using bio-inspired algorithms) to collect real-time soil moisture data across a field. The collected data can be used to create precise irrigation maps, optimizing water usage and reducing waste. This approach can be particularly beneficial in drought-prone regions.

7.3.7.4 Case Study 4

Bardsley England, a leading fruit producer, struggled with traditional methods for yield estimation and managing fruit variability within their orchards. These methods were inaccurate and time-consuming. Outfield Technologies deployed drones equipped with high-resolution cameras and AI software. These drones captured detailed images enabling 3D mapping of the trees for precise yield estimation. As a result, Bardsley England reported a 10% increase in productive yield. Additionally, the data-driven approach allowed for more precise resource allocation, optimizing costs and fruit quality.

7.3.7.5 Case Study 5

Traditional weed control methods in Thailand's durian industry involved manual spraying, leading to herbicide overuse and environmental damage. Farmers adopted DJI Agras T40 drones for weed control. These drones precisely apply herbicides, minimizing waste and environmental impact. The case study highlights a 50% reduction in herbicide use with drone spraying. Additionally, targeted application minimized risks to workers and surrounding ecosystems.

7.4 Real-World Applications

- **Crop yield prediction:** The FANETs can collect vast amounts of data on various parameters like soil health, plant growth, and weather conditions. This data can be fed into machine learning models to predict crop yields with greater accuracy, allowing farmers to optimize resource allocation and planning.

- **Frost protection:** Drones equipped with thermal imaging sensors can be deployed in a FANET to monitor field temperatures. Early detection of frost events can trigger irrigation systems to create a protective layer of ice on crops, preventing frost damage.
- **Livestock monitoring:** FANET-enabled drones can be used to monitor the health and location of livestock in large pastures. This data can be valuable for ensuring animal welfare and optimizing grazing management.
- **Crop monitoring and scouting:** High-resolution cameras and multispectral sensors mounted on drones allow farmers to capture detailed aerial images of their fields. This data is then analyzed using AI software to identify crop health issues like nutrient deficiencies, pest infestations, or diseases. Early detection enables farmers to take targeted interventions, minimizing crop loss and optimizing resource use.
- **Precision spraying:** Drones equipped with spray tanks can precisely apply fertilizers, pesticides, and herbicides. Unlike traditional methods (ground vehicles or airplanes), drones can **target specific areas** of the field, minimizing waste and environmental impact. This targeted approach also reduces risks to workers' health from exposure to chemicals.
- **Irrigation management:** Drones equipped with thermal imaging cameras can identify areas of a field with varying moisture levels. This data helps farmers optimize irrigation, ensuring crops receive the right amount of water and reducing water waste.
- **Field mapping and area measurement:** Drones can be used to create 3D maps of agricultural fields, providing valuable insights into field topography, drainage patterns, and overall size. This information is crucial for tasks like planning crop planting, irrigation systems, and optimizing land use.
- **Seed spreading and planting:** Some drones are equipped with mechanisms to spread seeds across large areas, particularly effective for reforestation projects or planting cover crops. Emerging technologies are exploring the possibility of using drones for precision planting of seeds in specific locations.

7.5 Conclusion

Bio-inspired routing protocols for FANETs hold immense potential for revolutionizing data collection and decision-making in PA. As research progresses, these algorithms are expected to become more efficient, robust, and better suited for real-world deployment, leading to a future of data-driven, efficient, and sustainable agricultural practices. By harnessing the power of FANETs and bio-inspired routing, agriculture has the potential to become significantly more efficient, sustainable, and data-driven. A precise application of resources will minimize environmental impact. Farmers will be empowered to make informed decisions based on real-time data insights.

References

1. Bitas, D., Papastavros, K., Tsipianitis, D., A Smart Cablebot for Spectral Imaging and Micro-climate measurements inside Greenhouses, in: *11th International Conference on Information, Intelligence, Systems and Applications, IISA 2020*, 2020, doi: 10.1109/IISA50023.2020.9284392.
2. Rangappa, N., Prasad, Y.R.V., Dubey, S.R., LEDNet: Deep Learning-Based Ground Sensor Data Monitoring System. *IEEE Sens. J.*, 22, 1, 842–850, 2022, doi: 10.1109/JSEN.2021.3129173.
3. Sharma, R.P., Ramesh, D., Pal, P., Tripathi, S., Kumar, C., IoT-Enabled IEEE 802.15.4 WSN Monitoring Infrastructure-Driven Fuzzy-Logic-Based Crop Pest Prediction. *IEEE Internet Things J.*, 9, 4, 3037–3045, 2022, doi: 10.1109/JIOT.2021.3094198.
4. Mohammad El-Basioni, B.M. and Abd El-Kader, S.M., Mission-based PTR triangle for multi-UAV systems flight planning. *Ad Hoc Netw.*, 142, 103115, 2023, doi: 10.1016/j.adhoc.2023.103115.
5. Pandey, C., Sethy, P.K., Behera, S.K., Vishwakarma, J., Tande, V., Smart agriculture: Technological advancements on agriculture-A systematical review, in: *Deep Learning for Sustainable Agriculture*, pp. 1–56, 2022, doi: 10.1016/B978-0-323-85214-2.00002-1.
6. Saif, S., Roy, P., Chowdhury, C., Biswas, S., Maulik, U., Smart e-agriculture monitoring systems, in: *AI, Edge and IoT-based Smart Agriculture*, pp. 183–203, 2021, doi: 10.1016/B978-0-12-823694-9.00002-5.
7. Sokullu, R., IoT in Agriculture: Irrigation Monitoring and Control System Example, in: *2020 7th International Conference on Energy Efficiency and Agricultural Engineering, EE and AE 2020 - Proceedings*, 2020, doi: 10.1109/EEAE49144.2020.9279037.

8. Gautam, A.K. and Yadav, R., Energy efficient hybrid routing protocol for Wireless Sensor Networks Using AI Technique. *EAI Endorsed Trans. Scalable Inf. Syst.*, 9, 35, 1–9, 2022, doi: 10.4108/eai.3-11-2021.171757.
9. Marszalek, M., Körner, M., Schmidhalter, U., Prediction of multi-year winter wheat yields at the field level with satellite and climatological data. *Comput. Electron. Agric.*, 194, 106777, 2022, doi: 10.1016/j.compag.2022.106777.
10. Joshi, P., Das, D., Udutalapally, V., Pradhan, M.K., Misra, S., RiceBioS: Identification of Biotic Stress in Rice Crops Using Edge-as-a-Service. *IEEE Sens. J.*, 22, 5, 4616–4624, 2022, doi: 10.1109/JSEN.2022.3143950.
11. Sankar, S., Srinivasan, P., Luhach, A.K., Somula, R., Chilamkurti, N., Energy-aware grid-based data aggregation scheme in routing protocol for agricultural internet of things. *Sustain. Comput. Inf. Syst.*, 28, 100422, 2020, doi: 10.1016/j.suscom.2020.100422.
12. Boursianis, A.D., *et al.*, Internet of Things (IoT) and Agricultural Unmanned Aerial Vehicles (UAVs) in smart farming: A comprehensive review. *Internet Things (Netherlands)*, 18, 100187, 2022, doi: 10.1016/j.iot.2020.100187.
13. Zhu, L. and Li, F., Agricultural data sharing and sustainable development of ecosystem based on block chain. *J. Clean. Prod.*, 315, 2021, doi: 10.1016/j.jclepro.2021.127869.
14. Liu, X., Zhou, Q., Deng, D., Opportunistic cooperative TDMA scheme for FANETs. *Phys. Commun.*, 36, 100745, 2019, doi: 10.1016/j.phycom.2019.100745.
15. Kumar, P. and Verma, S., Implementation of modified OLSR protocol in AANETs for UDP and TCP environment. *J. King Saud Univ. Comput. Inf. Sci.*, 34, 4, 1305–1311, 2022, doi: 10.1016/j.jksuci.2019.07.009.
16. Geetha, S.P., Sundari, N.M.S., Ramkumar, J., Karthikeyan, R., Energy Efficient Routing in Quantum Flying *Ad Hoc* Network (Q-Fanet) Using Mamdani Fuzzy Inference Enhanced Dijkstra 'S Algorithm (Mfi-Eda). *J. Theor. Appl. Inf. Technol.*, 102, 9, 3708–3724, 2024.
17. Swapna, M.P. and Ramkumar, J., Multiple Memory Image Instances Stratagem to Detect Fileless Malware BT - Advancements in Smart Computing and Information Security, in: *International Conference on Advancements in Smart Computing and Information Security (ASCIS-2023)*, S. Rajagopal, K. Popat, D. Meva, S. Bajeja (Eds.), pp. 131–140, Springer Nature Switzerland, Cham, 2024.
18. Yang, X., Han, B., Zhang, G., Zheng, P., Bai, J., Qin, D., NOMA-Assisted Routing Algorithm Design for UAV *Ad Hoc* Relay Networks. *IEEE Sens. J.*, 23, 3, 3296–3312, 2023, doi: 10.1109/JSEN.2022.3232636.
19. Ojha, N.K., Pandita, A., Ramkumar, J., Cyber security challenges and dark side of AI: Review and current status, in: *Demystifying the Dark Side of AI in Business*, pp. 117–137, 2024, doi: 10.4018/979-8-3693-0724-3.ch007.
20. Jaganathan, R. and Ramasamy, V., Performance modeling of bio-inspired routing protocols in Cognitive Radio *Ad Hoc* Network to reduce end-to-end delay. *Int. J. Intell. Eng. Syst.*, 12, 1, 221–231, 2019, doi: 10.22266/IJIES2019.0228.22.

21. Ramkumar, J., Vadivel, R., Narasimhan, B., Boopalan, S., Surendren, B., Gallant Ant Colony Optimized Machine Learning Framework (GACO-MLF) for Quality of Service Enhancement in Internet of Things-Based Public Cloud Networking, in: *Data Science and Communication. ICTDsC 2023*. Studies in Autonomic, Data-driven and Industrial Computing, J.M.R.S. Tavares, J.J.P.C. Rodrigues, D. Misra, D. Bhattacherjee (Eds.), pp. 425–438, Springer Nature Singapore, Singapore, 2024, doi: 10.1007/978-981-99-5435-3_30.
22. Ramkumar, J., Jeen Marseline, K.S., Medhunhashini, D.R., Relentless Firefly Optimization-Based Routing Protocol (RFORP) for Securing Fintech Data in IoT-Based *Ad-Hoc* Networks. *Int. J. Comput. Netw. Appl.*, 10, 4, 668–687, 2023, doi: 10.22247/ijcna/2023/223319.
23. Ramkumar, J. and Vadivel, R., CSIP—cuckoo search inspired protocol for routing in cognitive radio ad hoc networks, in: *Advances in Intelligent Systems and Computing*, pp. 145–153, Springer, Singapore, 2017, doi: 10.1007/978-981-10-3874-7_14.
24. Ramkumar, J. and Vadivel, R., Improved frog leap inspired protocol (IFLIP) – for routing in cognitive radio *ad hoc* networks (CRAHN). *World J. Eng.*, 15, 2, 306–311, 2018, doi: 10.1108/WJE-08-2017-0260.
25. Tsao, K.Y., Girdler, T., Vassilakis, V.G., A survey of cyber security threats and solutions for UAV communications and flying *ad-hoc* networks. *Ad Hoc Netw.*, 133, 102894, 2022, doi: 10.1016/j.adhoc.2022.102894.
26. Oubbati, O.S., Lakas, A., Zhou, F., Güneş, M., Lagraa, N., Yagoubi, M.B., Intelligent UAV-assisted routing protocol for urban VANETs. *Comput. Commun.*, 107, 93–111, 2017, doi: 10.1016/j.comcom.2017.04.001.
27. Jayaraj, D., Ramkumar, J., Lingaraj, M., Sureshkumar, B., AFSORP: Adaptive Fish Swarm Optimization-Based Routing Protocol for Mobility Enabled Wireless Sensor Network. *Int. J. Comput. Netw. Appl.*, 10, 1, 119–129, Jan. 2023, doi: 10.22247/ijcna/2023/218516.
28. Lingaraj, M., Sugumar, T.N., Felix, C.S., Ramkumar, J., Query aware routing protocol for mobility enabled wireless sensor network. *Int. J. Comput. Netw. Appl.*, 8, 3, 258–267, 2021, doi: 10.22247/ijcna/2021/209192.
29. Vadivel, R. and Ramkumar, J., QoS-enabled improved cuckoo search-inspired protocol (ICSIP) for IoT-based healthcare applications, in: *Inc. Internet Things Healthc. Appl. Wearable Devices*, pp. 109–121, 2019, doi: 10.4018/978-1-7998-1090-2.ch006.
30. Ramkumar, J. and Vadivel, R., Improved Wolf prey inspired protocol for routing in cognitive radio *Ad Hoc* networks. *Int. J. Comput. Netw. Appl.*, 7, 5, 126–136, 2020, doi: 10.22247/ijcna/2020/202977.
31. Sadek, R.A., Abd-alazeem, D.M., Abbassy, M.M., A New Energy-efficient Multi-hop Routing Protocol for Heterogeneous Wireless Sensor Networks. *Int. J. Adv. Comput. Sci. Appl.*, 12, 11, 481–491, 2021, doi: 10.14569/IJACSA.2021.0121154.

32. Jiang, J.L., Li, S.Y., Liao, M.L., Jiang, Y., Application in Disease Classification based on KPCA-IBA-LSSVM. *Procedia Comput. Sci.*, 154, 109–116, 2018, doi: 10.1016/j.procs.2019.06.017.
33. Kaur, K., Salgotra, R., Singh, U., An improved firefly algorithm for numerical optimization, in: *Proceedings of 2017 International Conference on Innovations in Information, Embedded and Communication Systems, ICIIECS 2017*, pp. 1–5, 2017, doi: 10.1109/ICIIECS.2017.8275914.
34. Ram, M.S., Rao, K., Krishna Rao, S., Trust based cluster head selection and secure routing in wireless sensor networks using Cat Swarm optimization and firefly algorithms. *J. Adv. Res. Dyn. Control Syst.*, 12, 2, 387–400, 2020, doi: 10.5373/JARDCS/V12I2/S20201028.
35. Olatinwo, S.O. and Joubert, T.H., Energy efficiency maximization in a wireless powered IoT sensor network for water quality monitoring. *Comput. Netw.*, 176, 107237, 2020, doi: 10.1016/j.comnet.2020.107237.
36. Singh, H., Tyagi, S., Kumar, P., Gill, S.S., Buyya, R., Metaheuristics for scheduling of heterogeneous tasks in cloud computing environments: Analysis, performance evaluation, and future directions. *Simul. Model. Pract. Theory*, 111, 102353, 2021, doi: 10.1016/j.simpat.2021.102353.
37. Karthiga, M., Santhi, V., Sountharrajan, S., Hybrid optimized convolutional neural network for efficient classification of ECG signals in healthcare monitoring. *Biomed. Signal Process. Control*, 76, 103731, 2022, doi: 10.1016/j.bspc.2022.103731.
38. Senthilkumar, A., Ramkumar, J., Lingaraj, M., Jayaraj, D., Sureshkumar, B., Minimizing Energy Consumption in Vehicular Sensor Networks Using Relentless Particle Swarm Optimization Routing. *Int. J. Comput. Netw. Appl.*, 10, 2, 217–230, 2023, doi: 10.22247/ijcna/2023/220737.
39. Jaganathan, R. and Vadivel, R., Intelligent Fish Swarm Inspired Protocol (IFSIP) for Dynamic Ideal Routing in Cognitive Radio *Ad-Hoc* Networks. *Int. J. Comput. Digit. Syst.*, 10, 1, 1063–1074, 2021, doi: 10.12785/ijcds/100196.
40. Menakadevi, P. and Ramkumar, J., Robust Optimization Based Extreme Learning Machine for Sentiment Analysis in Big Data. *2022 Int. Conf. Adv. Comput. Technol. Appl. ICACTA 2022*, Mar. 2022, pp. 1–5, doi: 10.1109/ICACTA54488.2022.9753203.
41. Ramkumar, J., Kumuthini, C., Narasimhan, B., Boopalan, S., Energy Consumption Minimization in Cognitive Radio Mobile *Ad-Hoc* Networks using Enriched *Ad-Hoc* On-demand Distance Vector Protocol, in: *2022 International Conference on Advanced Computing Technologies and Applications, ICACTA 2022*, 2022, doi: 10.1109/ICACTA54488.2022.9752899.
42. Ramkumar, J., Vadivel, R., Narasimhan, B., Constrained Cuckoo Search Optimization Based Protocol for Routing in Cloud Network. *Int. J. Comput. Netw. Appl.*, 8, 6, 795–803, 2021, doi: 10.22247/ijcna/2021/210727.
43. Mani, L., Arumugam, S., Jaganathan, R., Performance Enhancement of Wireless Sensor Network Using Feisty Particle Swarm Optimization

Protocol. *ACM Int. Conf. Proceeding Ser*, pp. 1–5, Dec. 2022, doi: 10.1145/3590837.3590907.

44. Ramkumar, J., Dinakaran, S.S., Lingaraj, M., Boopalan, S., Narasimhan, B., IoT-Based Kalman Filtering and Particle Swarm Optimization for Detecting Skin Lesion, in: *Lecture Notes in Electrical Engineering*, K. Murari, N. Prasad Padhy, S. Kamalasadan (Eds.), pp. 17–27, Springer Nature Singapore, Singapore, 2023, doi: 10.1007/978-981-19-8353-5_2.
45. Ramkumar, J. and Vadivel, R., Whale optimization routing protocol for minimizing energy consumption in cognitive radio wireless sensor network. *Int. J. Comput. Netw. Appl.*, 8, 4, 455–464, 2021, doi: 10.22247/ijcna/2021/209711.
46. Deitch, J., *Disruptive Fintech: The Coming Wave of Innovation in Financial Services with Thought Leadership Provided by CEOs*, De Gruyter, Boston, 2021, doi: 10.1515/9783110650471.
47. Lee, S.W., *et al.*, An Energy-Aware and Predictive Fuzzy Logic-Based Routing Scheme in Flying *Ad Hoc* Networks (FANETs). *IEEE Access*, 9, 129977–130005, 2021, doi: 10.1109/ACCESS.2021.3111444.

8

Disaster Recovery Management in FANETs

Amit Kumar[1]*, Sachin Ahuja[1] and Ganesh Gupta[2]

[1]*Dept. of Computer Science & Engineering, Chandigarh University, Mohali, Punjab, India*

[2]*Dept. of Computer Science & Engineering, Sharda University, Greater Noida, U.P., India*

Abstract

Flying *ad hoc* networks (FANETs) have emerged as a promising technology for a wide range of applications, including aerial surveillance, disaster relief, environmental monitoring, and communication in remote areas. However, operating in dynamic and unpredictable environments makes FANETs vulnerable to various disasters, such as communication link failures, node crashes, and natural calamities. Effective disaster recovery management is essential to ensure the continuity and reliability of FANET operations. This abstract presents an overview of disaster recovery management in FANET, outlining the challenges and potential solutions to mitigate the impact of disasters on network performance. We explore the unique characteristics of FANET, such as mobility, limited resources, and dynamic topology, which add complexity to the recovery process. The primary focus of disaster recovery management in FANETs is on three main aspects: fault detection, fault tolerance, and network reconfiguration. We discuss existing fault detection techniques, including node health monitoring, link quality assessment, and anomaly detection, to swiftly identify potential disruptions. Moreover, we investigate various fault tolerance mechanisms, such as redundancy-based approaches, cooperative communications, and load balancing, which aim to maintain network functionality even in the presence of failures. Furthermore, we examine network reconfiguration strategies, encompassing node relocation, route reestablishment, and adaptive protocols, to dynamically adjust FANET operations in response to

**Corresponding author*: amityadav038@gmail.com; ORCID: https://orcid.org/0009-0002-5708-3225

Sachin Ahuja: ORCID: https://orcid.org/0000-0002-2565-6685

Ganesh Gupta: ORCID: https://orcid.org/0000-0001-9623-292X

Taskeen Zaidi, Adarsh Kumar, Saurav Mallik and Keshav Kaushik (eds.) Autonomous Flying Ad-Hoc Networks, (165–176) © 2025 Scrivener Publishing LLC

disaster events. These strategies seek to optimize network performance, resource utilization, and resilience against disruptions. We also highlight the importance of communication and collaboration among nodes in disaster recovery scenarios, fostering the formation of multi-hop networks and facilitating the exchange of vital information. Additionally, leveraging cloud and edge computing resources for distributed data storage and processing can enhance disaster recovery capabilities. Lastly, we discuss research challenges and future directions for disaster recovery management in FANET, including energy-efficient recovery mechanisms, security considerations, and integration with ground-based networks. In conclusion, disaster recovery management is a critical aspect of ensuring the robustness and reliability of FANETs. By addressing the challenges posed by dynamic and unpredictable environments, as well as incorporating fault detection, fault tolerance, and network reconfiguration strategies, FANETs can be better prepared to handle disasters and maintain uninterrupted operations, making them invaluable for various critical applications.

***Keywords*:** Flying *ad hoc* networks (FANET), disaster recovery, fault detection, fault tolerance, network reconfiguration, node health monitoring, link quality assessment

8.1 Introduction

Flying *ad hoc* networks (FANETs) have emerged as a groundbreaking technology with a wide range of applications in both civilian and military domains. Comprised of unmanned aerial vehicles (UAVs) or drones that form a self-organizing network, FANETs offer unique capabilities for real-time data collection, surveillance, communication, and disaster response. However, operating in dynamic and challenging environments makes FANETs susceptible to various disruptions and failures, potentially jeopardizing mission-critical operations. Disasters in FANETs can arise from different sources, such as communication link interruptions, UAV malfunctions, unpredictable weather conditions, electromagnetic interference, and cyber-attacks. These disruptions can lead to communication breakdowns, loss of data, mission failure, and even safety hazards [2]. Effective disaster recovery management is indispensable to ensure the reliability and continuity of FANET operations.

Disaster recovery in FANETs encompasses a comprehensive set of strategies, protocols, and mechanisms aimed at minimizing the impact of disruptions and swiftly restoring normal network operations. The primary goal is to recover from unexpected events, maintain connectivity, and ensure the network's ability to adapt to dynamic changes while meeting the demands of critical applications. This paper explores the crucial aspects of disaster recovery in FANETs, addressing the challenges and proposing potential solutions to enhance the network's resilience and responsiveness. We delve into fault detection techniques to identify anomalies and failures in the network promptly. Fault tolerance mechanisms are investigated to design robust and adaptive solutions that can withstand and recover from disruptions effectively.

Additionally, network reconfiguration strategies are explored to enable dynamic adjustments to the network topology, routing, and resource allocation in the aftermath of disasters. These strategies aim to optimize network performance and ensure seamless data transfer and communication even in the face of adversity. Moreover, we examine the importance of communication and coordination among UAVs during disaster recovery scenarios. Cooperative networking approaches are explored to facilitate information exchange, task delegation, and collaborative decision-making among the network nodes. Furthermore, the integration of ground-based communication systems and other support infrastructure is considered to augment the capabilities of FANETs during recovery operations. The paper also highlights the significance of energy-efficient solutions, given the limited resources of UAVs, to prolong the network's endurance during disaster recovery efforts. Moreover, security considerations are discussed to safeguard the network against potential attacks and unauthorized access during vulnerable situations [3]. Disaster recovery in FANETs is a critical aspect of ensuring the network's reliability and effectiveness in dynamic and unpredictable environments. By addressing the challenges associated with fault detection, fault tolerance, network reconfiguration, and communication coordination, we can enhance a FANET's resilience and establish it as a valuable tool for disaster response, surveillance, and other critical applications. Through continued research and innovation, FANETs can contribute significantly to disaster management, humanitarian efforts, and various mission-critical scenarios in the future.

8.2 Related Work/Literature Survey

J. H. Lambert *et al.* [10] discussed the transportation system demands and system performance for critical disaster scenarios using a two-step methodology. The first is the number of trips to examine the situation and the behavioral aspects depending on the results of the survey. The second aspect is handing over these trips to the highway network by using the geographic information system. The content of the paper also measures the capacity ratio of link specific metrics like weighted average, speed, and traffic volume. The conclusion of this discussion is to estimate the number of evacuees using the network. The volume-capacity ratio can be improved.

Ahn *et al.* [12] described real-time rescue operations in critical disaster management. In this paper, the contents explain the realistic FANET environment simulation platform. The main purpose of this part is to design a stable UAV communication protocol with high rates of packet delivery and to bind the higher and lower limits of packet delivery without any kind of delay. The conclusion of this information comes with the development of a cluster-based flooding algorithm with the advantage of time division multiple access (TDMA). These kinds of two-level hierarchical clustering with real-time feasibility check methodology can improve the packet delivery ratio (PDR) and end-to-end delay. For future purposes, different clustering methodologies can be examined and can be customized for FANETs.

Dhall *et al.* [13] created a review of the protocols of different layers that are used to connect the nodes to exchange information when performing rescue operations in disaster areas. The focus of this knowledge is to provide the functions of protocols, including time synchronization and localization protocols on different layers such as the physical layer, data link layer, network layer, and application layer.

Pandey *et al.* [14] examined the limited battery power and high mobility of UAVs. In this paper, these domains will be reduced by using an efficient hybrid K-Means-Fruit Fly Optimization Clustering algorithm (KFFOCA). The performance and efficiency have been covered, and the results are analyzed against other optimization techniques like CLPSO, CACONET, GWOCNET, and ECRNET based on several performance parameters. The prominently covered contents are to increase the throughput, cluster-building time, end-to-end delay, and consumed energy.

8.3 Disaster Recovery Method of FANETs

Disaster recovery methods in FANETs are essential for maintaining network functionality and restoring communication in the face of disruptions or failures. The pictorial implementation of proposed disaster recovery methods in FANETs is:

These methods encompass a range of strategies and protocols designed to address the unique challenges posed by a FANET's dynamic and mobile nature. Here are some key disaster recovery methods employed in FANET:

1. Fault detection and localization: Effective fault detection mechanisms are crucial for identifying disruptions promptly.

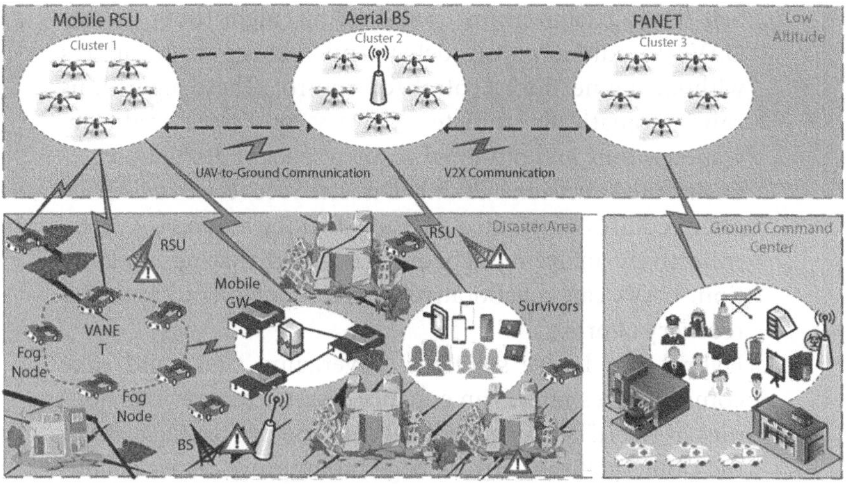

Figure 8.1 Disaster recovery in FANETs.

Various techniques, such as node health monitoring, link quality assessment, and anomaly detection, can be utilized to identify failing UAVs or links in the network [4].

2. Fault tolerance and redundancy: Fault tolerance methods aim to design FANETs with a redundancy to ensure continued operationa even if some nodes or links fail. Redundancy-based approaches involve creating backup paths or establishing redundant connections to maintain network connectivity.
3. Cooperative communications: Cooperative communication techniques enable UAVs to collaborate and relay data for nodes that have lost direct connectivity. By forming multi-hop routes, cooperative communications can enhance the chances of successful data delivery even in the presence of disruptions.
4. Load balancing: Load balancing strategies distribute network traffic and data processing tasks across UAVs to prevent overloading and improve overall network performance. In the event of node failures, load balancing can help redistribute tasks to other available nodes.
5. Dynamic reconfiguration: Network reconfiguration involves dynamically adjusting the network topology and routing protocols in response to disaster events. This process may include node relocation, route reestablishment, and adaptive routing protocols to adapt to changing network conditions [5].

6. Self-healing algorithms: Self-healing algorithms enable UAVs to autonomously detect and recover from failures without the need for centralized control. These algorithms can automatically initiate recovery processes, reducing response time and minimizing the impact of disruptions.
7. Distributed decision-making: Disaster recovery in a FANET often requires distributed decision-making among UAVs. By collectively analyzing network status and sharing information, UAVs can collaboratively make decisions to optimize recovery efforts.
8. Hybrid cloud-edge solutions: Leveraging cloud and edge computing resources can enhance disaster recovery capabilities in FANETs. Storing critical data and processing tasks in the cloud or at the network edge can improve data availability and reduce the burden on individual UAVs.
9. Energy-efficient recovery: Energy efficiency is crucial in FANETs due to limited UAV battery life. Recovery methods should prioritize energy conservation while ensuring successful recovery and network restoration.
10. Security considerations: Disaster recovery methods must address security concerns to safeguard the network from potential attacks or unauthorized access during vulnerable situations. Implementing secure communication protocols and access control mechanisms is essential.

Disaster recovery methods in FANETs play a vital role in ensuring network resilience and continuous operations in dynamic and challenging environments [6]. By combining fault detection, fault tolerance, network reconfiguration, and cooperative communication strategies, FANETs can effectively recover from disasters and fulfill their intended missions, including disaster response, surveillance, and communication in remote or critical areas.

8.4 Proposed New Solutions for Improved Disaster Management in FANETs

Improved methods for FANETs to manage disasters are essential to enhance the network's resilience, responsiveness, and effectiveness during critical situations. These methods aim to address the unique challenges of FANETs and optimize the disaster recovery process. Here are some advanced techniques that can improve disaster handling in FANETs:

1. Machine learning for fault prediction: Utilizing machine learning algorithms, such as neural networks and decision trees, can enable UAVs in the network to predict and preemptively handle potential faults. By learning from historical data and network behavior, UAVs can take proactive measures to avoid disruptions and optimize resource allocation.
2. Software-defined networking (SDN): Implementing SDN principles in FANETs allows for centralized network control and dynamic reconfiguration. A centralized controller can efficiently manage network resources and traffic flows, making it easier to adapt to changing conditions during disasters.
3. Mobility-aware routing: Developing mobility-aware routing protocols can improve data delivery and reduce latency in highly dynamic FANET scenarios. These protocols consider UAV mobility patterns to optimize routing decisions and ensure data reaches its destination effectively [7].
4. Hierarchical network organization: Introducing hierarchical network organizations can enhance disaster handling in FANETs. Dividing the network into multiple tiers with different roles (e.g., data aggregators, relay nodes, and central coordinators) can improve coordination, fault tolerance, and resource management.
5. Cognitive networking: Cognitive networking allows UAVs in FANETs to learn and adapt to changing environmental conditions and user demands. The UAVs can intelligently adjust their behavior and optimize network parameters during and after disasters, enhancing network performance.
6. Cross-layer optimization: Coordinating multiple layers of the communication protocol stack (e.g., physical, MAC, and routing layers) can improve disaster recovery performance. Cross-layer optimization enables efficient resource utilization and reduces the impact of disruptions on the network.
7. Energy harvesting and energy sharing: Integrating energy harvesting technologies into UAVs can prolong their endurance and improve their ability to handle disasters [8]. Additionally, energy sharing mechanisms among UAVs can help redistribute energy resources during prolonged recovery operations.

8. Hybrid communication technologies: Employing a combination of different communication technologies, such as satellite links, ground-based relays, and other UAVs, can enhance communication resilience in disaster scenarios. This hybrid approach ensures data delivery even when certain links or nodes are affected.
9. Social-aware networking: Leveraging social behavior patterns of UAVs and incorporating social-aware networking techniques can improve communication efficiency and cooperation during disaster recovery efforts.
10. Multi-objective optimization: Disaster handling in FANETs often involves multiple conflicting objectives, such as minimizing energy consumption, maximizing data delivery, and optimizing latency [9]. Multi-objective optimization algorithms can balance these objectives to achieve a well-rounded disaster recovery performance.

By integrating these advanced techniques into FANET disaster recovery management, the network can better adapt to dynamic and unpredictable conditions, minimize downtime, optimize resource utilization, and ultimately enhance its effectiveness in critical scenarios such as disaster response and relief operations. Continued research and innovation in these areas are crucial to realizing the full potential of FANETs in disaster handling and humanitarian efforts.

8.5 Conclusion

In conclusion, disaster recovery in FANETs is of paramount importance to ensure the reliability, resilience, and effectiveness of this dynamic and versatile technology in challenging and unpredictable environments. A FANET, comprising UAVs forming a self-organizing network, has shown great potential for critical applications, including disaster response, surveillance, communication, and environmental monitoring. However, the mobility of UAVs, limited resources, and dynamic topology make FANETs susceptible to various disruptions and failures. To address these challenges and optimize disaster recovery in FANETs, a range of innovative methods and techniques have been explored. Fault detection mechanisms, such as node health monitoring, link quality assessment, and anomaly detection, facilitate the swift identification of disruptions, enabling prompt recovery actions. Fault tolerance strategies, including redundancy-based approaches, cooperative communications, and load balancing, ensure the continuous operation of the network even in the presence of failures. Network reconfiguration techniques, such as dynamic topology adjustment and adaptive routing protocols, allow FANETs to adapt to changing conditions and maintain connectivity during and after disasters. Moreover, the integration of cloud and edge computing resources enhances disaster recovery capabilities by distributing data storage and processing tasks across the network. To further advance disaster recovery in FANETs, cutting-edge methods have been proposed, including machine learning for fault prediction, SDN, mobility-aware routing, and cognitive networking. These techniques empower UAVs to make informed decisions, optimize network parameters, and improve coordination during critical situations. Additionally, hierarchical network organizations, cross-layer optimization, energy harvesting, and hybrid communication technologies contribute to the network's resilience and communication efficiency in disaster scenarios. Social-aware networking fosters collaboration among UAVs, enhancing their cooperative behavior during recovery efforts. In summary, the continued research and implementation of these advanced methods for disaster recovery in FANETs are crucial to maximizing the network's potential in addressing critical challenges, including disaster response, relief operations, and surveillance missions. By proactively handling disruptions, optimizing resource utilization, and ensuring rapid recovery, a FANET can prove to be a valuable and indispensable technology for disaster management, humanitarian aid, and various mission-critical applications in the future. As technology continues to evolve, the potential for disaster recovery in FANETs will continue to expand, offering promising opportunities to improve disaster management and responses in a rapidly changing world.

8.6 Future Techniques for Disaster Recovery in FANETs

The future of disaster recovery in FANETs holds great promise as advancements in technology and research continue to address existing challenges and open new possibilities. As the field of FANETs evolve, several key trends and developments are expected to shape the future of disaster recovery:

Autonomy and artificial intelligence (AI): The integration of AI and autonomy into FANETs will play a significant role in disaster recovery. The UAVs equipped with advanced AI algorithms will be capable of making real-time decisions, coordinating recovery efforts, and adapting to dynamic and complex disaster scenarios without human intervention [10].

Swarm intelligence: A FANET's potential can be further enhanced by employing swarm intelligence techniques. By enabling UAVs to work collaboratively as a cohesive unit, swarm intelligence can optimize disaster recovery operations, improve data collection, and ensure reliable communication during critical situations.

The fifth generation of wireless technology and beyond: The deployment of the fifth generation of wireless technology and future communication technologies will greatly influence disaster recovery in FANETs. Faster and more reliable communication will facilitate real-time data exchange, enhance situational awareness, and enable a more effective disaster response and coordination [11].

Edge computing and Internet of Things integration: Integrating edge computing capabilities into FANETs will enable faster data processing and decision-making directly at the network's edge. Additionally, incorporating Internet-of-Things devices into the network will provide valuable sensor data, enhancing disaster detection and recovery planning.

Seamless integration with ground networks: Future FANETs will seamlessly integrate with ground-based communication systems to provide a holistic approach to disaster recovery. This integration will ensure continuous connectivity, efficient data sharing, and coordinated efforts between FANETs and ground-based responders [12, 13].

Privacy and security enhancements: As FANETs become more prevalent in disaster response and recovery, ensuring the privacy and security of sensitive data will be of utmost importance. Advanced encryption techniques, secure communication protocols, and robust authentication mechanisms will be vital in safeguarding data during disaster recovery operations [14].

Standardization and interoperability: The development of standardized protocols and interfaces for FANETs will promote interoperability among different UAVs and communication systems. This will enable seamless collaboration among heterogeneous UAVs and enhance disaster recovery coordination across various platforms.

Energy-efficiency and sustainability: Future FANETs will emphasize energy-efficient solutions to extend UAV endurance during disaster recovery missions. Advancements in energy harvesting technologies and improved battery life will further support sustained operations in critical scenarios [15].

Real-world testing and validation: The future of disaster recovery in FANETs will involve extensive real-world testing and validation. Large-scale simulations and field trials will be conducted to assess the effectiveness and scalability of new disaster recovery methods.

The future of disaster recovery in FANETs holds exciting prospects, driven by advancements in AI, communication technologies, and integration with ground networks. As researchers and developers continue to innovate and refine disaster recovery methods, the FANET will emerge as a dependable and resilient solution for disaster management, humanitarian aid, and critical missions, making a significant and positive impact on society and saving lives in times of crisis.

Bibliography

1. Faraci, G., Rizzo, S.A., Schembra, G., Green Edge Intelligence for Smart Management of a FANET in Disaster-Recovery Scenarios. *IEEE Trans. Veh. Technol.*, 72, 3, 3819–3831, 2023.
2. Ahmed, S.B.M., Hussain, S.A., Latiff, L.A., Ahmad, N., Sam, S.M., Performance Evaluation of FANET Routing Protocols in Disaster Scenarios, in: *2021 IEEE Symposium On Future Telecommunication Technologies (SOFTT)*, 2021, December, IEEE, pp. 46–51.
3. Basu, S., Banerjee, S., Halder, T., Das, A.K., Sarkar, S., Basak, A., Chakravarty, D., A Comparative Study on Propagation Models for Routing Protocols in FANETs, in: *2021 IEEE International India Geoscience and Remote Sensing Symposium (InGARSS)*, 2021, December, IEEE, pp. 528–531.
4. Ahmed, H., Bakr, M., Talib, M.A., Abbas, S., Nasir, Q., Unmanned aerial vehicles (UAVs) and artificial intelligence (AI) in fire related disaster recovery: analytical survey study, in: *2022 International Conference on Business Analytics for Technology and Security (ICBATS)*, 2022, February, IEEE, pp. 1–6.
5. Khan, M.A., Qureshi, I.M., Khanzada, F., A hybrid communication scheme for efficient and low-cost deployment of future flying *ad-hoc* network (FANET). *Drones*, 3, 1, 16, 2019.
6. Singal, G., Laxmi, V., Gaur, M.S., Rao, D.V., Kushwaha, R., UAVs reliable transmission for multicast protocols in FANETs, in: *2019 Sixth International Conference on Internet of Things: Systems, Management and Security (IOTSMS)*, 2019, October, IEEE, pp. 130–135.
7. Tropea, M., Fazio, P., De Rango, F., Cordeschi, N., A new fanet simulator for managing drone networks and providing dynamic connectivity. *Electronics*, 9, 4, 543, 2020.
8. Xu, J., Zeng, Y., Zhang, R., UAV-Enabled Wireless Power Transfer: Trajectory Design and Energy Optimization. *IEEE Trans. Wirel. Commun.*, 17, 8, 5092–5106, Aug. 2018.
9. Cheng, F., *et al.*, UAV Trajectory Optimization for Data Offloading at the Edge of Multiple Cells. *IEEE Trans. Veh. Technol.*, 67, 7, 6732–36, July 2018.

10. Lambert, J.H., Parlak, A.I., Zhou, Q., Miller, J.S., Fontaine, M.D., Guterbock, T.M., *et al.*, Understanding and managing disaster evacuation on a transportation network. *Accid. Anal. Prev.*, 50, 645–658, 2013, https://doi.org/10.1016/j.aap.2012.06.015.
11. Sun, G., Qin, D., Lan, T., Ma, L., Research on Clustering Routing Protocol Based on Improved PSO in FANET. *IEEE Sens. J.*, 21, 23, 27168–27185, 1 Dec.1, 2021, doi: 10.1109/JSEN.2021.3117496.
12. Ahn, T., Seok, J., Lee, I., Han, J., Reliable Flying IoT Networks for UAV Disaster Rescue Operations. *Mobile Inf. Syst.*, 2018, 1–12, 2018, doi: 10.1155/2018/2572460.
13. Dhall, R. and Dhongdi, S., Review of Protocol Stack Development of Flying *Ad-hoc* Networks for Disaster Monitoring Applications. *Arch. Comput. Methods Eng.*, 30, 37–68, 2023, https://doi.org/10.1007/s11831-022-09791-y.
14. Pandey, A., Shukla, P. K., Agrawal, R., An adaptive Flying *Ad-hoc* Network (FANET) for disaster response operations to improve quality of service (QoS). *Mod. Phys. Lett. B.*, 34, 10, 2050010, 2020.
15. Kim, G.H., Mahmud, I., Cho, Y.Z., Self-recovery scheme using neighbor information for multi-drone *ad hoc* networks, in: *2017 23rd Asia-Pacific Conference on Communications (APCC)*, 2017, December, IEEE, pp. 1–5.

9

AI-Based Cybersecurity Opportunities and Issues on the IIoT

Akashdeep Bhardwaj

Centre for Cybersecurity, School of Computer Science, UPES, Dehradun, India

Abstract

The rise of the Industrial Internet of Things (IIoT) has brought numerous benefits to industries, including improved efficiency, automation, and cost savings. Critical infrastructure security and safety are at risk, though, as new cybersecurity issues have also been brought about by it. One solution to these challenges is the use of artificial intelligence (AI) in cybersecurity. This chapter examines the possible advantages and disadvantages of using AI in IIoT cybersecurity. The writers talk about how AI can be utilized to improve the security of IIoT systems by assisting with threat, anomaly, and cyberattack identification. This chapter also discusses some potential drawbacks that may arise from using AI in cybersecurity, including the potential for false positives or negatives, the requirement for ongoing model monitoring and updating, and the vulnerability of AI to manipulation by hostile parties. Providing a thorough grasp of the possible advantages and difficulties of AI-based cybersecurity for IIoT is the goal of this chapter, which also offers suggestions for how these technologies might be used to improve the security of vital infrastructure.

Keywords: AI, cybersecurity, IoT, industrial IoT, cyberthreats, cyber issues

9.1 Introduction

Businesses are now able to optimize production processes, cut costs, and increase operational efficiency thanks to the Industrial Internet of Things (IIoT) [1], which has completely changed the way industries function.

Email: bhrdwh@yahoo.com

Taskeen Zaidi, Adarsh Kumar, Saurav Mallik and Keshav Kaushik (eds.) Autonomous Flying Ad-Hoc Networks, (177–198) © 2025 Scrivener Publishing LLC

The development of "smart factories," where equipment, sensors, and machines are linked to a centralized network to allow real-time data collection and analysis, has been made possible by the IIoT. Although the IIoT has benefited several businesses, it has also brought forth new difficulties, notably about cybersecurity. The use of IIoT devices has dramatically increased the attack surface of critical infrastructure. In traditional manufacturing systems, the process was closed, with minimal connectivity between the machines. The IIoT systems, on the other hand, provide additional risks as their gadgets are networked to the internet and may contact external networks. As a result, cyberattacks on IIoT systems have increased significantly in recent years, with attackers leveraging a range of techniques such as malware, social engineering, and phishing attacks to infiltrate industrial control systems (ICSs) [2].

To counter the increasing threat of cyberattacks, a range of technologies and techniques have been developed, including firewalls, intrusion detection systems, and encryption. However, these technologies are not always effective in preventing sophisticated attacks. The application of artificial intelligence (AI) in cybersecurity is therefore gaining popularity [3]. Because AI makes it possible for businesses to identify and react to cyberattacks instantly, it has the potential to completely change the cybersecurity environment. Artificial intelligence systems are capable of spotting patterns and abnormalities in vast amounts of data that might point to an active assault. Artificial intelligence may also be used to anticipate assaults by spotting possible weaknesses in past data and evaluating it.

This chapter aims to explore the application of AI in IIoT cybersecurity. We will discuss the benefits and limitations of AI-based cybersecurity, and how it can be used to strengthen the security of IIoT systems. We will also address some of the potential issues associated with the implementation of AI in cybersecurity, such as the possibility of false positives or negatives, the need for constant monitoring and updating of the AI model, and the potential for AI to be manipulated by malicious actors. As the IIoT continues to grow, so do the potential risks associated with cyberattacks. The IIoT devices are connected to the internet and are accessible from anywhere in the world, creating new vulnerabilities that can be exploited by attackers. These devices are often located in remote or hard-to-reach locations, making them difficult to monitor and secure. The IIoT systems often rely on legacy systems and technologies that were not designed with cybersecurity in mind. This creates a challenge for organizations as they try to secure these systems while maintaining their functionality. Consequently, there is an increasing demand for fresh perspectives on cybersecurity that can tackle the difficulties presented by the IIoT.

Cybersecurity powered by AI has become a viable answer to these problems. From past data, AI systems may learn to recognize trends and abnormalities that can lead to the presence of an attack. As a result, the period between an assault and an organization's reaction to a cyberattack might decrease. Artificial intelligence is also useful for anticipating future attacks and identifying possible weak points. There are additional difficulties when using AI in cybersecurity. For example, AI models must be constantly monitored and updated to ensure they remain effective, as attackers may seek to evade detection by changing their tactics. Additionally, there is a risk that AI models may produce false positives or negatives, leading to unnecessary alerts or missed attacks. Despite these challenges, the potential benefits of AI-based cybersecurity for IIoT are significant. Artificial intelligence can help organizations detect and respond to cyberattacks in real-time, reducing the risk of downtime and data loss. It can also help organizations identify potential vulnerabilities and take proactive steps to prevent future attacks.

In this chapter, we will explore the potential of AI-based cybersecurity for IIoT, including its benefits and limitations. We will also discuss some of the key issues associated with the implementation of AI in cybersecurity and offer insights into how these technologies can be effectively deployed to enhance the security of critical infrastructure. Finally, we will provide a roadmap for future research in this field.

This chapter provides a comprehensive understanding of the potential benefits and challenges of AI-based cybersecurity for IIoT and offers insights into how these technologies can be effectively deployed to enhance the security of critical infrastructure. The chapter is organized as follows: In Section 9.2, we provide a brief overview of the IIoT, its benefits, and cybersecurity challenges. Section 9.3 discusses the role of AI in cybersecurity, including its potential benefits and limitations. In Section 9.4, we explore the application of AI in IIoT cybersecurity and discuss its potential impact. In Section 9.5, we address some of the potential issues associated with the implementation of AI in cybersecurity. Finally, in Section 9.6, we provide a conclusion and discuss the future research directions in this field.

9.2 Application of AI in IIoT Cybersecurity

Artificial intelligence has various applications in IIoT cybersecurity [4], from detecting and responding to cyberattacks to predicting and preventing future threats. Here are some of the key applications of AI in IIoT cybersecurity; some of these points may overlap as the overall domain is within cybersecurity:

9.2.1 Anomaly Detection

The identification of anomalies is among the most important uses of AI in IIoT cybersecurity [5]. Real-time analysis of massive data sets by AI algorithms can spot trends and abnormalities that could point to a cyberattack. As a result, there is less chance of downtime and data loss and enterprises can swiftly identify and respond to threats. Artificial intelligence can assist with IIoT anomaly detection in cybersecurity by instantly evaluating massive amounts of data and seeing trends and abnormalities that can point to a cyberattack. Here is how AI may be applied to IIoT cybersecurity anomaly detection:

- Data analysis: To find trends and abnormalities, AI systems may examine massive amounts of data, such as system logs, network traffic data, and user activity data. Artificial intelligence systems can learn what typical behavior looks like through analyzing past data, and they can recognize variations from this behavior that can point to an assault.
- Machine learning: Machine learning techniques may also be used by AI systems to find trends and abnormalities in data. Artificial intelligence systems can recognize novel trends and abnormalities that might point to a possible assault by learning from past data.
- Real-time detection: AI systems can identify abnormalities instantly, giving companies the ability to react swiftly to possible cyberattacks. Artificial intelligence systems can detect possible risks by evaluating data in real time and sending notifications to security professionals so they can respond right away.
- Predictive analytics: Predictive analytics is another tool that AI systems may use to see possible risks before they become serious. Artificial intelligence systems can anticipate possible dangers and take proactive steps to avert assaults by examining patterns and trends in data.
- Pattern recognition: To detect such risks, AI systems can also make use of pattern recognition techniques. Artificial intelligence systems can recognize novel hazards that may not have been observed previously by examining patterns in data.
- Automated response: Automation of procedures for responding to possible cyberattacks is another application

for AI algorithms. Organizations may lower the risk of data loss and downtime by automating response procedures, which enable them to respond to potential threats more swiftly and effectively.

Thus, by evaluating massive amounts of data in real-time, applying machine learning algorithms to find patterns and abnormalities, and automating response procedures, AI may aid in anomaly detection for IIoT in cybersecurity. Organizations may safeguard their vital infrastructure and lower the risk of data loss and downtime by promptly identifying and addressing any cyberattacks.

9.2.2 Threat Intelligence

Artificial intelligence may also be used to collect and evaluate data related to threat intelligence [6], which helps firms remain ahead of possible adversaries. Artificial intelligence systems can recognize patterns and trends in past data that can point to potential future assaults. By using this knowledge, preventative measures against future assaults can be developed. Artificial intelligence can assist with cybersecurity threat intelligence for IIoT by collecting and evaluating massive amounts of data from several sources to spot possible risks. Artificial intelligence may be applied to threat intelligence in IIoT cybersecurity in the following ways:

- Data collection: To learn more about possible risks, AI algorithms may acquire data from a variety of sources, including social media, network traffic, and system logs. Using this data, patterns and trends that can point to a possible assault might be found.
- Machine learning: AI systems may detect patterns and trends in data that can point to a possible assault by using machine learning techniques. Artificial intelligence systems can recognize novel risks that may not have been observed previously by learning from past data.
- Real-time analysis: Real-time data analysis by AI systems helps businesses react swiftly to possible risks. Artificial intelligence systems can detect possible risks by evaluating data in real-time and sending notifications to security professionals so they can respond right away.
- Predictive analytics: Predictive analytics is a tool that AI systems may use to find possible dangers before they become

serious. Artificial intelligence systems can participate possible dangers and take proactive steps to avert assaults by examining patterns and trends in data.

- Threat modeling: Threat modeling approaches may be used by AI algorithms to find vulnerabilities and possible attack scenarios in IIoT systems. By identifying potential attack scenarios, organizations can develop proactive measures to prevent attacks and protect their critical infrastructure.
- Automated response: Automating procedures for responding to possible cyberattacks is possible using AI algorithms. Organizations may lower the risk of data loss and downtime by automating response procedures, which enable them to respond to potential threats more swiftly and effectively.

Thus, by collecting and evaluating vast amounts of data, applying machine learning algorithms to spot patterns and trends, and automating reaction procedures, AI can provide threat intelligence for IIoT in cybersecurity. Organizations may mitigate the risk of data loss and downtime and safeguard their vital infrastructure by proactively preventing attacks and recognizing possible risks.

9.2.3 Network Security

Artificial intelligence may also be used to detect possible threats and weaknesses, improving network security [7]. Artificial intelligence algorithms can detect possible security threats like malware infections and unauthorized access attempts by examining network traffic data. By using this information, preventative actions against attackers can be taken in advance. Artificial intelligence can help in network security for IIoT in cybersecurity by detecting and preventing cyberattacks on industrial networks. Here are some ways where AI can be used for network security in IIoT:

- Intrusion detection: Network traffic data may be analyzed by AI algorithms to find possible IIoT system breaches or assaults. Artificial intelligence algorithms can recognize such attacks and notify security personnel to act by studying network data in real-time.
- Behavior analysis: Industrial systems' typical behavior may be learned by AI algorithms, which can then recognize unusual or suspicious behavior that can point to a cyberattack. Artificial intelligence algorithms can recognize

patterns of activity that can point to a danger by examining system logs and network traffic data.

- Threat intelligence: Machine learning techniques may be employed by AI algorithms to evaluate threat intelligence data and detect possible dangers to IIoT systems. Through the collection of data from several sources, including open-source intelligence, social media, and the dark web, AI algorithms can recognize possible dangers and proactively avert assaults.
- Vulnerability assessment: Automated vulnerability evaluations of IIoT systems may be carried out using AI algorithms to find possible security flaws. Artificial intelligence algorithms can detect vulnerabilities that might be used by hackers by examining software applications and network setups.
- Real-time response: AI algorithms have the capability to react instantly to possible cyberattacks by preventing harmful traffic, cutting off network connections, or isolating compromised computers. Organizations may lessen the effects of cyberattacks and avoid data loss and disruption by taking immediate action.
- Security automation: Security procedures like patch management, system upgrades, and user access control may be automated with the application of AI algorithms. Organizations may lower the possibility of human mistake and guarantee that security regulations are followed consistently by automating security procedures.

Artificial intelligence can help in network security for IIoT in cybersecurity by detecting and preventing cyberattacks, analyzing behavior data, performing vulnerability assessments, and automating security processes. Organizations may lower the risk of data loss and downtime, defend their vital infrastructure from cyberattacks, and increase network security by implementing AI.

9.2.4 User Behavioral Analysis

Artificial intelligence is also useful for identifying possible security threats and analyzing user behavior [8]. Artificial intelligence systems are capable of spotting anomalous behavior that might point to a possible attack by examining user activity data. Investigating any security incidents and taking necessary action may be done with the use of this information.

By examining the behavior of industrial systems and recognizing unusual or suspect activity that can point to a cyberattack, AI can assist in behavioral analysis for IIoT cybersecurity. The following are some methods that AI may be applied to behavioral analysis in IIoT cybersecurity:

- Learning normal behavior: AI algorithms may analyze past data from system logs and network traffic to understand the typical behavior of industrial systems. Artificial intelligence algorithms can recognize unusual or suspicious activity that might point to a cyberattack by learning about the typical behavior of systems.
- Real-time monitoring: AI algorithms can track IIoT system behavior in real time by examining network traffic data and system logs to spot unusual or suspicious activity. Artificial intelligence algorithms that monitor systems in real-time are able to identify such risks immediately and notify security professionals to take appropriate action.
- Anomaly detection: Machine learning techniques can be employed by AI algorithms to identify deviations in system behavior that could potentially point to a cyberattack. AI systems can identify previously unseen attack types by examining patterns and trends in data.
- Predictive analytics: Predictive analytics is a tool that AI systems may use to find possible dangers before they become serious. Artificial intelligence systems can anticipate possible dangers and take proactive steps to avert assaults by examining patterns and trends in data.
- User behavior analysis: AI systems are capable of analyzing user behavior data to spot potentially harmful or dangerous threats. Artificial intelligence systems can recognize unusual activity that might be a sign of a cyberattack by examining user access records and network traffic data.
- Behavioral profiling: By using AI algorithms, security teams may more readily spot possible vulnerabilities by creating behavioral profiles of IIoT systems and their users. Artificial intelligence algorithms may assist security teams in promptly identifying and taking necessary action in response to suspicious or aberrant conduct by generating behavioral profiles.

Thus, AI can help in behavioral analysis for IIoT in cybersecurity by learning normal behavior, monitoring systems in real-time, detecting anomalies, using predictive analytics, analyzing user behavior, and creating behavioral profiles. Organizations may lower the risk of data loss and downtime by employing AI for behavioral analysis to identify possible risks more rapidly and take proactive steps to avoid assaults.

9.2.5 Predictive Maintenance

Artificial intelligence may also be used to forecast IIoT device maintenance requirements [9]. Artificial intelligence systems can detect possible problems before they become serious by evaluating device data. It is possible to schedule maintenance tasks and avoid unscheduled downtime by using this information. Artificial intelligence uses machine learning algorithms to evaluate data from industrial systems and identify probable equipment failures or security risks. This helps with predictive maintenance for IIoT in cybersecurity. The following are some ways that AI may be applied to IIoT cybersecurity predictive maintenance:

- Predictive analytics: Artificial intelligence algorithms can predict future equipment failures or security threats by analyzing data from industrial systems, including sensors, logs, and network traffic. By analyzing patterns and trends in data, AI algorithms can identify potential issues before they become critical, enabling maintenance teams to take proactive measures to prevent downtime or security breaches.
- Condition monitoring: AI algorithms can use machine learning techniques to monitor the condition of industrial equipment and identify potential issues. By analyzing data from sensors and other monitoring devices, AI algorithms can detect changes in equipment conditions that may indicate a potential failure.
- Root cause analysis: AI algorithms can analyze data from industrial systems to identify the root cause of equipment failures or security incidents. By analyzing system logs and network traffic data, AI algorithms can identify the source of the problem and take appropriate action to prevent similar incidents in the future.
- Predictive maintenance scheduling: Predictive analytics may be used by AI algorithms to plan maintenance tasks according to the anticipated condition of industrial machinery. Organizations may decrease equipment failure risk and downtime by scheduling maintenance activities at the best times.

- Security incident prediction: AI systems can forecast future security events, such cyberattacks or data breaches, by using machine learning techniques. Artificial intelligence algorithms can identify possible attacks and preventing security issues by proactively evaluating threat information feeds and network traffic data.
- Security automation: Patch management, system upgrades, and user access control are just a few of the security procedures that AI algorithms can automate. Organizations may guarantee that security regulations are constantly implemented and lower the possibility of human mistake by automating security operations.

Overall, AI can help in predictive maintenance for IIoT in cybersecurity by predicting potential equipment failures or security issues, monitoring equipment conditions, analyzing root causes, scheduling maintenance activities, predicting security incidents, and automating security processes. By using AI for predictive maintenance, organizations can reduce downtime, prevent equipment failures, and protect their critical infrastructure from cyberthreats.

9.2.6 Fraud Detection

Artificial intelligence has the potential to identify and stop fraud in IIoT systems [9]. Artificial intelligence systems are capable of spotting unusual activity that could be a sign of fraud by examining transaction data. This information can be used to prevent financial losses and protect sensitive data. Artificial intelligence can help in fraud detection for IIoT in cybersecurity by analyzing data from industrial systems and identifying anomalies or suspicious activity that may indicate fraudulent behavior. Here are some ways AI can be used for fraud detection in IIoT cybersecurity:

- Anomaly detection: AI algorithms can use machine learning techniques to detect anomalies in data from industrial systems, such as sensor data, logs, and network traffic. By analyzing patterns and trends in data, AI algorithms can identify abnormal behavior that may indicate fraudulent activity.
- Real-time monitoring: AI algorithms can monitor industrial systems in real-time, analyzing system logs and network traffic data to identify suspicious activity. By monitoring

systems in real time, AI algorithms can quickly detect potential fraud and alert security teams to act.

- Predictive analytics: Predictive analytics is a tool that AI systems may use to spot any fraudulent activity before it happens. Artificial intelligence systems can anticipate possible fraud and take proactive steps to avoid it by evaluating past data and behavioral trends.
- User behavior analysis: AI systems can analyze user behavior data to spot fraudulent activities or insider threats. Artificial intelligence systems can recognize unusual activity that can point to fraud by examining user access records and network traffic data.
- Fraudulent pattern recognition: AI systems can identify patterns of fraudulent activity using machine learning approaches, such as anomalous transactions or behavioral shifts. Artificial intelligence systems can recognize patterns of behavior that point to fraud by examining past data.
- Fraud prevention: Real-time transaction monitoring and risk scoring are two examples of fraud protection procedures that may be automated with AI algorithms. Organizations may lower the possibility of human mistake and guarantee that fraudulent conduct is identified and stopped by automating these procedures.

Hence, AI can help in fraud detection for IIoT in cybersecurity by using anomaly detection, real-time monitoring, predictive analytics, user behavior analysis, fraudulent pattern recognition, and fraud prevention. Organizations may lower their risk of financial loss, safeguard their brand, and guarantee the integrity of their business processes by utilizing AI for fraud detection.

9.2.7 Cybersecurity Automation

Cybersecurity procedures like incident response and threat remediation may also be automated using AI [10]. Organizations may respond to cyberattacks more swiftly and effectively by automating these procedures, which lowers the chance of data loss and downtime. By automating tedious and time-consuming security duties, AI can assist in cybersecurity automation for IIoT, freeing up security staff to concentrate on more intricate and strategic responsibilities. Here are some ways AI can be used for cybersecurity automation in IIoT:

- Threat detection and response: Real-time threat detection and industrial system monitoring are both possible with AI algorithms. Organizations can minimize the risk of harm to critical infrastructure by promptly identifying and responding to cyberthreats through the automation of threat detection and response processes.
- Vulnerability management: Industrial system vulnerabilities may be found, and remedial efforts can be prioritized using AI algorithms. Organizations may make sure that serious vulnerabilities are fixed quickly by automating vulnerability management.
- Incident response: Processes involved in incident response, such incident triage, containment, and remediation, may be automated with the help of AI algorithms. Organizations may reduce the effect of cyberattacks and respond to security issues more rapidly by automating an incident response.
- Compliance management: Processes in compliance management, such as reporting and policy enforcement, may be automated with AI algorithms. Organizations may lower the possibility of human mistake and guarantee that compliance standards are constantly followed by automating compliance management.
- User behavior analytics: AI algorithms may be used to track user activity and spot any insider threats. Organizations may minimize the risk of data loss or system penetration by promptly identifying and responding to insider threats through the automation of user behavior analytics.
- Security orchestration and automation: Security procedures can be automated, and security actions may be coordinated across many platforms and systems, thanks to AI algorithms. Organizations may enhance their security posture and mitigate the possibility of human mistakes by using security orchestration and automation.

Therefore, by automating threat detection and response, vulnerability management, incident response, compliance management, user behavior analytics, and security orchestration and automation, AI can aid in cybersecurity automation for IIoT. Organizations may strengthen their security posture, lower their risk of cyberattacks, and boost the effectiveness of their security operations by automating cybersecurity processes with AI.

All things considered, AI has the power to revolutionize IIoT cybersecurity by empowering enterprises to anticipate possible risks, take proactive steps to thwart attacks, and quickly identify and address cyberattacks. Even while integrating AI into IIoT cybersecurity presents certain difficulties, there are a lot of potential advantages, which makes AI a vital instrument for safeguarding vital infrastructure.

9.3 Potential Benefits and Issues

The applications of AI in cybersecurity are many and beneficial. By enhancing threat detection, automating security processes, boosting productivity, improving incident response, offering better threat intelligence, facilitating better user behavior analysis, and enhancing risk management, AI has the potential to improve an organization's cybersecurity capabilities. Artificial Intelligence can help businesses uncover risks that may have gone undetected, automate a lot of repetitive and time-consuming security processes, and free security professionals to work on more complicated and strategic duties. Furthermore, compared to conventional approaches, AI algorithms may identify possible cyberthreats more rapidly and correctly by analyzing vast amounts of data from different sources. All things considered, incorporating AI into cybersecurity operations may strengthen an organization's defenses against cyberattacks, lower their likelihood, and boost the effectiveness of those efforts. Artificial intelligence in cybersecurity has a number of possible advantages, including:

- Improved threat detection: AI systems can identify possible cyberthreats faster and more precisely than traditional approaches by analyzing enormous amounts of data from many sources. By doing this, firms may identify hazards that could have gone undetected and react to possible assaults faster.
- Enhanced security automation: Numerous time-consuming and repetitive security operations, including handling incidents, keeping an eye out for vulnerabilities, and identifying and addressing threats, may be automated by AI. Security staff may be able to concentrate on more intricate and strategic tasks as a result.
- Increased efficiency: AI may help firms increase the effectiveness of their security operations and lower the possibility of human mistake, which will save money, by automating security jobs.

- Improved incident response: By automating incident triage and response procedures, AI algorithms may assist businesses in responding to security issues more swiftly and efficiently. By doing this, the organization's exposure to cyberattacks may be reduced.
- Enhanced threat intelligence: To detect possible threats and forecast upcoming assaults, AI algorithms may be used to evaluate threat intelligence data from a variety of sources, including internal and external data sources.
- Better user behavior analysis: AI can analyze user activity records and network traffic data to assist enterprises monitor user behavior and identify any insider threats.
- More accurate risk management: AI algorithms may be used to evaluate the entire security risk posture of a business and offer suggestions for risk management tactics.

In general, AI can greatly improve an organization's cybersecurity capabilities through a variety of means, including better threat detection, enhanced security automation, increased efficiency, improved incident response, improved threat intelligence, improved user behavior analysis, and improved risk management.

Additionally, there are some possible concerns and challenges with the integration of AI in IIoT cybersecurity. The opaqueness of AI decision-making processes is one major obstacle. Artificial intelligence's sophisticated algorithms may be hard to comprehend and may not give concise justifications for its judgments. Organizations may find it difficult to recognize and resolve such problems or prejudices as a result. Furthermore, inadequate understanding of AI might pose a serious threat to IIoT cybersecurity. Artificial intelligence systems could be unable to identify fresh or developing dangers, leaving businesses open to intrusion. Artificial intelligence algorithms may be biased as a result of the data they are trained on, which might result in unfair or inaccurate judgments. This raises further concerns. Cyberattacks on AI systems are also a growing concern, as attackers may target these systems to compromise their effectiveness and gain access to sensitive data. Finally, the implementation of AI in IIoT cybersecurity can be complex and expensive, requiring significant investment in hardware, software, and training. To prevent AI from being hacked, organizations must be aware of these possible issues and take action to resolve them. Some of these actions include ensuring diversity in training data, retaining human oversight, and installing suitable security measures to defend AI from cyberattacks. While AI has many advantages for cybersecurity, there are also some drawbacks and concerns that businesses need to be aware of, such as:

- Lack of transparency: Organizations may find it difficult to recognize and rectify any problems or biases in AI algorithms due to their complexity and lack of clear explanations of their decision-making processes.
- Bias: AI algorithms may be biased based on the data they are trained on, which can lead to incorrect or discriminatory decisions.
- Limited knowledge: AI algorithms may not be able to recognize new or emerging threats, which can leave organizations vulnerable to attack.
- Limited human oversight: The reliance on AI for cybersecurity may reduce the need for human intervention, but it can also limit the ability of security teams to intervene in critical situations.
- Cyberattacks on AI: As AI is used more extensively in cybersecurity, it may become a target for cyberattacks, which can compromise its effectiveness and lead to data breaches.
- Complexity: The implementation of AI in cybersecurity can be complex and require specialized skills and resources, which can be challenging for organizations with limited expertise or resources.
- Cost: Implementing AI in cybersecurity can be expensive, requiring significant investment in hardware, software, and training.

Thus, organizations should be aware of the potential problems and issues of AI in cybersecurity and take steps to address them, such as providing transparency in AI decision-making, ensuring diversity in training data, maintaining human oversight, and implementing appropriate security measures to protect AI from cyberattacks.

9.4 AI-Cybersecurity Use Cases

There are many real-world use-case examples of how AI is being used for IIoT cybersecurity. Here are some notable ones:

9.4.1 Siemens

The German multinational company Siemens uses AI to protect its ICSs against cyberattacks. The business's AI system, known as Charter of Trust [11], analyzes network traffic using machine learning algorithms to look for irregularities that could point to a cyberattack. Additionally, the system makes recommendations for mitigation techniques and prospective hazards by utilizing predictive analytics. As a founding member of the worldwide project known as the Charter of Trust, Siemens is committed to advancing cybersecurity and data protection. Siemens has created a number of AI-based cybersecurity solutions as part of this effort.

Siemens provides real-time threat detection and response capabilities using AI in its Charter of Trust products. The business has created an AI-based security analytics platform that analyzes network traffic and detects possible security risks using machine learning techniques. The platform can recognize vulnerabilities, spot unusual network behavior and trends, and offer remedial suggestions. Siemens also uses AI in its industrial anomaly detection solution, which is designed to protect ICSs against cyberattacks. The solution uses machine learning algorithms to monitor ICSs and detect abnormal behavior, such as unauthorized changes to system configurations or unexpected changes in production processes. Security teams may be immediately notified by the system of such dangers, enabling them to investigate and take appropriate action. Siemens has created a number of cybersecurity services in addition to its AI-based solutions, all aimed at assisting businesses in strengthening their cybersecurity posture. Threat intelligence, incident response, and cybersecurity assessments are some of these services. Siemens is therefore utilizing AI to improve cybersecurity and offer real-time threat detection and response capabilities. Siemens can monitor networks, identify irregularities, and take prompt, efficient action in response to any threats by utilizing machine learning and artificial intelligence. This helps safeguard vital infrastructure and other systems from cyberattacks.

9.4.2 Honeywell

The US-based conglomerate Honeywell has developed an AI-based cybersecurity platform called Honeywell Forge Cybersecurity Suite [12]. This tool monitors network traffic and looks for irregularities that can point to a cyberattack using machine learning techniques. In addition, the system can create warnings for possible threats and doing risk assessments.

Honeywell Forge Cybersecurity Suite is an AI-based cybersecurity platform developed by the US-based conglomerate Honeywell. The platform is designed to provide comprehensive protection against cyberattacks on ICSs, which are commonly used in critical infrastructure such as power plants, oil and gas refineries, and manufacturing facilities. Using cutting-edge AI and machine learning algorithms, the Honeywell Forge Cybersecurity Suite monitors network traffic and looks for irregularities that could point to a cyberattack. Real-time threat identification is possible with this technology, which can also provide warnings for security experts to look at. Additionally, it offers risk analyses and suggestions for mitigating techniques to deal with vulnerabilities that have been found.

The Honeywell Forge Cybersecurity Suite's capability to continuously monitor ICSs is one of its primary characteristics. To give a thorough picture of the organization's cybersecurity posture, the platform analyzes data from many sources, such as network traffic, user activity, and system logs, using sophisticated AI algorithms. Security analysts may monitor and handle cybersecurity issues with the help of a consolidated dashboard offered by the Honeywell Forge Cybersecurity Suite. Security teams can swiftly examine vast amounts of security data and spot any problems thanks to the platform's powerful analytics features. The Honeywell Forge Cybersecurity Suite has sophisticated threat intelligence and incident response capabilities in addition to AI-based capabilities. The platform is integrated with Honeywell's Managed Security Services, which offer round-the-clock cyberattack detection and response. Using AI and machine learning, the Honeywell Forge Cybersecurity Suite is an extensive and sophisticated cybersecurity platform that offers ICSs continuous monitoring, threat detection, and incident response capabilities. It is a great illustration of how AI may be applied to strengthen critical infrastructure protection against cyberattacks.

9.4.3 Darktrace

Artificial intelligence algorithms are used by UK-based cybersecurity business Darktrace [13] to quickly identify and address cyberthreats. The Enterprise Immune System, an AI-based system, uses machine learning to examine network traffic and spot possible threats. To improve its defenses against potential threats in the future, the system can also incorporate insights from previous attacks. Darktrace offers a variety of AI-driven Internet of Things (IoT) cybersecurity solutions. Unsupervised machine learning, the foundation of the company's novel approach to cybersecurity, enables the system to identify and react to cyberthreats in real-time without the need for human interaction. The AI algorithms at Darktrace operate by identifying typical patterns of behavior inside an enterprise's IoT network. The system can swiftly identify unusual behavior, which might be a sign of a cyberattack, once it has learned what is typical. Unusual traffic patterns, illegal access attempts, and other questionable activities that conventional cybersecurity solutions could overlook are examples of this.

The Enterprise Immune System, an AI-based platform that can identify and address cyberthreats throughout an organization's whole network, is one of Darktrace's primary offerings. The system analyzes data from cloud services, IoT devices, and other sources using machine learning algorithms, and when it notices unusual activity, it instantly sends out notifications. Additionally, Darktrace provides an Industrial Immune System that is intended especially for IoT networks in industrial settings, such energy grids and manufacturing facilities. The Industrial Immune System can identify cyberthreats that could be directed at ICSs by using AI algorithms to track and examine the behavior of IoT devices. Moreover, automatic response features that can neutralize threats instantly are part of Darktrace's products. To stop malware from spreading further, the system can automatically quarantine infected devices or block suspicious network activity.

Because of this, Darktrace's AI algorithms allow the business to offer IoT networks real-time threat detection and response capabilities, assisting enterprises in thwarting cyberattacks and safeguarding their vital assets. Darktrace can offer a highly effective cybersecurity solution that is well-suited to the dynamic and ever-evolving world of IoT since it is continually learning about and reacting to new threats.

9.4.4 Symantec

Artificial intelligence is used by US-based cybersecurity business Symantec to identify and stop cyberattacks on its clients' networks. Machine learning techniques are used by its AI-based Symantec Endpoint Protection system [14] to examine network data and spot any threats. Even if a danger has never been encountered before, the system can recognize it and react against it. Symantec employs AI to quickly identify and neutralize online threats. The business offers AI-powered solutions that are intended to monitor and analyze massive amounts of data from various sources, including cloud services, traditional IT infrastructure, and IoT devices.

The integrated cyber defense platform, which leverages machine learning algorithms to evaluate data from an organization's entire network, is one of Symantec's primary AI-based products. To stop more harm, the platform may automatically isolate compromised devices or block shady network activity. It is built to recognize and react to cyberthreats in real-time. The endpoint detection and response solution from Symantec, which is intended to identify and counteract sophisticated threats on endpoints including laptops, desktops, and servers, is likewise powered by the company's AI algorithms. The endpoint detection and response system analyzes endpoint behavior and looks for irregularities that could be signs of a cyberattack using machine learning techniques. The system may automatically neutralize a threat by quarantining compromised endpoints or obstructing malicious traffic as soon as it is identified. Symantec's threat intelligence service, which gives businesses up-to-date knowledge on new cyberthreats, is also powered by AI algorithms. The service analyzes massive amounts of data from many internet sources, such as hacker forums, social media, and other sources, using machine learning. When the system notices a possible threat, it then notifies the companies so they may take preventative measures to safeguard their assets.

All things considered, Symantec's usage of AI algorithms allows the business to offer incredibly efficient cybersecurity solutions that can identify and address cyberthreats instantly. Through constant learning and adaptation to emerging threats, Symantec maintains an advantage over hackers and offers its clients the best security available.

9.4.5 IBM

Artificial intelligence is used by IBM's Watson [15] for cybersecurity to recognize and address cyberthreats. Massive volumes of security data are analyzed by the system using machine learning to find any dangers. Watson may also direct security analysts in event response and offer suggestions for mitigating techniques. These real-world use case studies exhibit AI's ability to strengthen ICSs' defenses against cyberattacks and show how successful it is in IIoT cybersecurity. Artificial intelligence is employed by IBM to identify and address online risks. Massive volumes of data from a variety of sources, including IoT devices, cloud services, and conventional IT infrastructure, may be monitored and analyzed by IBM's AI-based solutions.

The Watson for Cyber Security platform, which uses machine learning algorithms to evaluate security data from an organization's network, is one of IBM's primary AI-based products. To stop more harm, the platform may automatically isolate compromised devices or block shady network activity. It is built to recognize and react to cyberthreats in real time. IBM's The X-Force Threat Intelligence service, which gives businesses up-to-date knowledge on new cyberthreats, is also powered by AI algorithms.

The service analyzes massive amounts of data from many internet sources, such as hacker forums, social media, and other sources, using machine learning. When the system notices a possible threat, it then notifies the companies so they may take preventative measures to safeguard their assets.

Among IBM's AI-powered products is the IBM Security QRadar Advisor with Watson, a tool that helps security analysts make better judgments by analyzing security events using machine learning. The technology can identify trends and abnormalities that might point to a cyberattack and can offer suggestions for fixing them. IBM's AI-based solutions can also assist companies in automating security procedures so they can react to cyberattacks faster. By automating repetitive security operations with machine learning, the company's Security Operations Center (SOC) automation solution frees up analysts to concentrate on more sophisticated threats. All things considered, IBM's usage of AI algorithms allows the business to offer extremely effective cybersecurity solutions that can quickly identify and address online threats. Through constant learning and adaptation to emerging threats, IBM maintains a competitive edge against hackers and offers its clients the best security available.

The main points that readers would anticipate seeing in a thorough examination of this subject are covered here, including the potential advantages and restrictions of AI-based cybersecurity, the different ways that AI can be used to improve IIoT security, and some of the issues that must be resolved to guarantee the efficacy of these technologies. To those who are curious about the possible advantages of these technologies and how to use them to improve the security of critical infrastructure, providing insights into the real-world applications of AI in IIoT cybersecurity would be very helpful. Additionally, the writers are offering a nuanced and balanced viewpoint on this complicated subject by pointing out some of the possible problems and difficulties related to the application of AI in cybersecurity. This chapter makes a significant addition to the subject of IIoT cybersecurity by giving readers a thorough grasp of the role that AI may play in safeguarding critical infrastructure as well as the issues that must be resolved to guarantee these technologies' efficacy.

9.5 Conclusion

In summary, the use of AI in IIoT cybersecurity has great potential to improve the security of vital infrastructure. Artificial intelligence enables enterprises to proactively minimize possible hazards before they cause harm by being able to recognize threats, abnormalities, and cyberattacks. Additionally, businesses may employ AI to automate a variety of cybersecurity procedures, freeing up security professionals to concentrate on more important tasks. Nevertheless, the application of AI in cybersecurity is not without its difficulties. These include the possibility of false positives or negatives, the requirement for ongoing monitoring and upgrading of AI models to guarantee their efficacy, and the danger of AI being exploited by bad actors. The advantages of AI in IIoT cybersecurity are evident despite these obstacles. Artificial intelligence has the potential to be a useful tool for improving the security of vital infrastructure if the proper procedures and precautions are followed. Organizations must keep up with the most recent advancements in AI-based cybersecurity as the threat landscape changes and strive to create strong and practical plans for utilizing these technologies to safeguard their operations and assets.

References

1. What is IIoT (Industrial Internet of Things)? - Definition from, https://www.techtarget.com/iotagenda/definition/Industrial-Internet-of-Things-IIoT (accessed: Mar. 16, 2023).
2. Industrial Control System - Definition, https://www.trendmicro.com/vinfo/us/security/definition/industrial-control-system (accessed: Feb. 07, 2023).
3. Artificial Intelligence (AI) for Cybersecurity, IBM, https://www.ibm.com/security/artificial-intelligence (accessed: Feb. 25, 2023).
4. Business Guide to Industrial IoT (Industrial Internet of Things), https://www.i-scoop.eu/internet-of-things-iot/industrial-internet-things-iiot-saving-costs-innovation/ (accessed: Feb. 25, 2023).
5. What is Anomaly Detection?, Anodot, https://www.anodot.com/blog/what-is-anomaly-detection/ (accessed: Jan. 25, 2023).
6. What Is Threat Intelligence?, Recorded Future, https://www.recordedfuture.com/threat-intelligence (accessed: Jan. 07, 2023).
7. Industrial Internet of Things (IIoT) Security: Everything You Need to, https://www.imperosoftware.com/blog/industrial-internet-of-things-complete-guide/ (accessed: Feb. 16, 2023).
8. What Is User and Entity Behavior Analytics (UEBA)?, https://www.rapid7.com/fundamentals/user-behavior-analytics/ (accessed: Mar. 25, 2023).
9. The Role of Big Data Analytics in Industry 4.0, https://blog.rgbsi.com/big-data-analytics-in-industry-4.0 (accessed: Feb. 07, 2023).
10. Topics Understanding automation What is security automation?, https://www.redhat.com/en/topics/automation/what-is-security-automation (accessed: Feb. 25, 2023).

11. Cybersecurity - Topic Areas - Global, https://www.siemens.com/global/en/company/topic-areas/cybersecurity.html (accessed: Feb. 16, 2023).
12. OT Cybersecurity Solutions, Honeywell Forge, https://www.honeywellforge.ai/us/en/industries/cybersecurity (accessed: Mar. 25, 2023).
13. Darktrace AI: Combining Supervised and Unsupervised Machine, https://darktrace.com/resources/darktrace-ai-combining-supervised-and-unsupervised-machine-learning (accessed: Feb. 07, 2023).
14. Symantec unveils AI-powered endpoint security, TechNative, https://technative.io/symantec-unveils-ai-powered-endpoint-security/ (accessed: Mar. 25, 2023).
15. Artificial Intelligence (AI) for Cybersecurity, IBM, https://www.ibm.com/security/artificial-intelligence (accessed: Jan. 07, 2023).

10

Exploring the Synergy of Fog Computing and FANETs for Next-Generation Networking

Akashdeep Bhardwaj

Center for Cybersecurity, School of Computer Science, UPES, Dehradun, India

Abstract

The combination of flying *ad-hoc* networks (FANETs) with fog computing appears to be a potential way to transform next-generation networking as technology advances. The synergy between fog computing and FANETs is examined in this chapter, along with their distinct qualities and the possible advantages and difficulties of integrating them. Fog computing enables real-time processing, lower latency, and increased efficiency by bringing networking, storage, and computing closer to the network's edge. Conversely, a FANET is an autonomous, dynamic network of unmanned aerial vehicles (UAVs) that interact with one another in a decentralized fashion. By combining the strengths of fog computing and FANETs, new opportunities arise for applications requiring mobility, scalability, and distributed processing. This chapter delves into the technical aspects of integrating fog computing and FANETs, including communication protocols, resource management, and data processing strategies. It explores the challenges faced in achieving seamless integration, such as network connectivity, security, and coordination among UAVs. Moreover, it examines the potential applications and use cases where the fusion of fog computing and FANETs can bring significant advantages, such as disaster management, surveillance, precision agriculture, and smart cities.

***Keywords*:** Fog computing, flying *ad hoc* networks, FANET, next-generation networking, synergy, integration, distributed processing

Email: bhrdwh@yahoo.com

Taskeen Zaidi, Adarsh Kumar, Saurav Mallik and Keshav Kaushik (eds.) Autonomous Flying Ad-Hoc Networks, (199–232) © 2025 Scrivener Publishing LLC

10.1 Introduction

A decentralized computing infrastructure called fog computing brings cloud computing capabilities to the edge of the network. Fog computing [1] brings data processing and storage closer to the data source, allowing for real-time processing and lower latency than traditional cloud computing, which mostly houses these functions in distant data centers. The idea that computing infrastructure should reside near the surface, much like a fog, is referred to as "fog" [2]. One of fog computing's main features is its closeness to data sources; this technology reduces latency by processing data near its point of generation or consumption. For applications that require real-time or near real-time computing, this closeness makes quicker reaction times possible. Next, there is the distributed architecture [3]. Fog computing makes use of this design, which distributes computing resources across a variety of edge devices, such as switches, routers, gateways, and Internet of Things (IoT) devices. Data processing may be done more efficiently thanks to this distribution, which also lessens the need to send data to a distant cloud server. Scalability is another factor, as fog computing infrastructure is made to grow to accommodate a large number of devices and manage applications that require a lot of data. It facilitates load balancing and dynamic resource allocation to guarantee effective resource use. Numerous technologies, such as embedded systems, cellphones, and IoT devices, are supported by fog computing [4]. The IoT ecosystem's increasing number of connected devices is accommodated by this heterogeneity, which also allows for a variety of applications [5]. Fog computing lowers the dangers involved with sending sensitive data to a distant cloud server by processing data locally at the edge. It improves security and privacy by enclosing sensitive data in a more regulated and restricted setting.

Flying *ad hoc* networks [6] are self-organizing networks of unmanned aerial aircraft that interact with one another independently. They constitute a dynamic network where UAVs serve as both data providers and relays, rather than depending on a permanent infrastructure like base stations or access points [7]. The key aspects of a FANET include autonomy and self-organization as FANETs enable UAVs [9] to operate autonomously and self-organize into a network [8]. The UAVs establish communication links with neighboring drones and collaboratively maintain network connectivity without centralized control. The dynamic network topology of a FANET in nature is due to the movement of UAVs. The network topology changes continuously as drones move, enter, or leave the network. Flying *ad hoc* networks are designed to adapt to these dynamic changes and maintain connectivity. The UAVs' mobility and flexibility [10] in FANETs provide the ability to fly and move freely, making them suitable for various applications requiring mobility and flexibility. They can be deployed in remote or inaccessible areas, perform aerial surveillance, support search and rescue operations, and more. Flying *ad hoc* networks enable UAVs to act as data collectors and relays. They can collect data from sensors or cameras on board and transmit it to other UAVs or ground stations. Applications including environmental monitoring, disaster relief, and surveillance can benefit from these capabilities. Flying *ad hoc* networks promote collaborative decision-making among UAVs. Drones can share information, coordinate their actions, and collectively optimize network performance, routing decisions, and resource utilization.

By integrating fog computing and FANETs, the strengths of both technologies can be leveraged to create a powerful network infrastructure. The combination enables real-time processing, reduced latency, mobility, and distributed computing capabilities, opening new possibilities for applications that require edge intelligence, efficient data processing, and seamless connectivity in dynamic environments. This integration of fog computing and FANETs is motivated by several factors that highlight the potential benefits and synergies between these two technologies. Let us explore the motivations in detail:

- Reduced latency: The goal of fog computing is to reduce latency by moving processing power closer to the network's edge. By integrating fog computing with FANETs, which enables communication and data exchange among UAVs in a decentralized manner, processing and decision-making can occur closer to the data source. This reduces the latency associated with transmitting data to a remote cloud server for processing, enabling real-time or near real-time applications that require immediate response times.
- Mobility and flexibility: FANETs leverage UAVs that can move freely in the airspace, making them highly suitable for applications requiring mobility and flexibility. By integrating fog computing with FANETs, computational resources and services can be made available on-demand wherever the UAVs are present. This allows for dynamic deployment of services and enables UAVs to access edge computing capabilities while on the move. For example, UAVs conducting surveillance operations can offload computational tasks to nearby fog nodes, enhancing their operational efficiency.

- Edge intelligence and data processing: Intelligent processing and analytics at the network's edge, nearer the data source, are made possible *via* fog computing. By integrating fog computing with FANETs, UAVs can benefit from localized data processing capabilities. The UAVs can perform initial data filtering, analysis, and decision-making at the edge, reducing the need to transmit large volumes of data to a remote cloud for processing. By keeping critical data on the local network, this not only lowers bandwidth costs but also improves privacy and security.
- Resilience and redundancy: FANETs exhibit self-organizing and resilient characteristics, as UAVs can dynamically form and maintain communication links even in the absence of a fixed infrastructure. Replicated fog nodes may be placed throughout the network by incorporating fog computing, guaranteeing the availability of processing resources even in the case of UAV malfunctions or network outages. This redundancy improves the system's overall resilience and fault tolerance, enabling data processing and ongoing operations in harsh settings.
- Scalability and network capacity: Fog computing and FANETs both offer scalability in their respective domains. Fog computing allows for the efficient distribution of computational resources across multiple edge devices, enabling scalable and parallel processing of data. Flying *ad hoc* networks can scale by adding more UAVs to the network, allowing for increased network capacity and improved data transmission capabilities. The scalability of both technologies may be utilized by combining fog computing with FANETs, allowing for the effective processing of massive data volumes and supporting the network's expanding number of connected devices.
- Mission-critical applications: The integration of FANET with fog computing creates opportunities for mission-critical applications across a range of industries. For example, in disaster management scenarios, UAVs equipped with sensors can collect and process critical data in real-time using fog computing capabilities, aiding in rapid decision-making and response coordination. Similar to this, fog computing combined with FANETs can offer sophisticated

video analytics, object recognition, and tracking at the edge, lowering bandwidth needs and improving system responsiveness in surveillance and monitoring applications.

Therefore, the goal of utilizing the advantages of both technologies to provide real-time processing, decreased latency, mobility, edge intelligence, scalability, and resilience in dynamic networking settings is what drives the combination of fog computing with FANETs. Numerous applications, such as those needing prompt decision-making, effective data processing, and dependable communication under difficult situations, are made possible by this integration.

When it comes to combining fog computing and FANETs, next-generation networking [11] is essential. An uninterrupted connection between the FANET's UAVs and the fog computing infrastructure is guaranteed *via* next-generation networking. It provides [12] robust and efficient communication protocols, routing mechanisms, and network management techniques to support the dynamic and self-organizing nature of FANET. The UAVs may easily access fog computing resources and services because to this connection, which facilitates effective data processing and decision-making. Fog computing and FANET resource allocation and usage are optimized *via* next-generation networking. To guarantee the effective use of computing resources, network bandwidth, and energy in the integrated system, fog computing and FANET resource allocation provide intelligent resource management, load balancing, and job scheduling approaches. By using resources more effectively, the network's overall performance, scalability, and responsiveness are improved, increasing the potential of FANET and fog computing.

Edge intelligence and processing power in FANETs and fog computing are made possible by next-generation networking. At the network's edge, in closer proximity to the data source, it permits the implementation of intelligent algorithms, machine learning models, and analytics. Real-time data analysis, decision-making, and insight production are made possible by this edge intelligence in FANET applications. Next-generation networking gives UAVs the ability to analyze data locally, lowering latency, speeding reaction times, and facilitating prompt decision-making by combining fog computing with FANETs.

Next-generation networking addresses the security and privacy concerns associated with integrating FANETs and fog computing. To secure data transport, storage, and processing in the integrated system, next-generation networking includes strong security methods, authentication protocols, encryption techniques, and access control systems. By taking these precautions, sensitive data security and privacy are improved, protecting the confidentiality and integrity of data in FANETs and fog computing applications. Next-generation networking provides scalable and future-proof architecture for the integrated FANET and fog computing system. It can adapt to the expanding amount of data, the growing number of connected devices, and the changing needs of applications. The design makes it possible to seamlessly connect new technologies with established ones, such edge computing, 5G networks, and the IoT. This guarantees that the integrated system may grow and change to meet the needs of new applications and technical developments.

Next-generation networking enhances the user experience in FANET and fog computing applications. It enables low-latency communication, real-time data processing, and seamless connectivity, resulting in improved response times, higher data throughput, and a more interactive and immersive user experience. This improved user experience is vital for applications such as immersive augmented reality, collaborative UAV operations, remote monitoring, and real-time situational awareness.

Thus, next-generation networking plays a critical role in integrating FANET and fog computing by ensuring seamless connectivity, efficient resource utilization, edge intelligence, security, scalability, and enhanced user experience. It provides the foundational infrastructure and technologies necessary to leverage the synergies between FANET and fog computing, enabling the development of advanced applications and unlocking the full potential of these technologies in dynamic networking environments.

The effects of fog computing-integrated FANETs on the general design of next-generation networks are covered in this chapter. It explores how this integration can enhance network resilience, improve data processing efficiency, and enable novel services and applications. The considerations for deploying fog computing and FANETs in practical scenarios are also examined, including regulatory aspects, energy constraints, and infrastructure requirements. The synergy between fog computing and FANETs is thoroughly explored in this chapter, which also offers insights into the difficulties and opportunities associated with combining these technologies for next-generation networking. It is an invaluable tool for scholars, professionals, and decision-makers who want to comprehend the possibilities of this new paradigm and how it may affect network designs in the future.

10.2 Fog Computing and FANET

By extending cloud computing capabilities to the network's edge, fog computing is a decentralized computing infrastructure. By moving networking, storage, and computing closer to the data source, it seeks to overcome the drawbacks of conventional cloud computing and enable real-time processing while lowering latency and boosting system performance. Reduced latency, increased efficiency, scalability, edge intelligence, and greater privacy and security are just a few advantages of fog computing, which moves processing, storage, and networking capabilities closer to the network edges. Businesses may support a variety of applications that need quick insights, low-latency connectivity, and an effective use of computational resources by utilizing fog computing to provide real-time processing, improve the performance of edge devices, and support these objectives.

Self-organizing networks of UAVs that communicate with one another decentralized are FANETs. They function as a dynamic network where UAVs serve as both data sources and relays, independent of permanent infrastructure like base stations or access points. Flying *ad hoc* networks have drawn a lot of interest because of their potential for use in a variety of applications, such as communication in inaccessible or distant places, environmental monitoring, disaster response, and surveillance. They enable UAVs to operate autonomously and self-organize into a network without relying on centralized control. Each UAV can establish communication links with neighboring drones, exchange information, and collaborate to maintain network connectivity. This decentralized nature allows FANETs to adapt to dynamic environments and operate in scenarios where fixed infrastructure is not available or feasible. Flying *ad hoc* networks exhibit a dynamic network topology due to the movement of UAVs. The network structure continuously changes as drones move, enter, or leave the network. They employ dynamic routing protocols and algorithms to maintain connectivity and adapt to the changing network topology. This dynamic characteristic is particularly valuable in scenarios where UAVs need to explore and cover a large area or operate in challenging and unpredictable environments. Flying *ad hoc* networks leverage the mobility and flexibility of UAVs. The UAVs can fly and move freely in three-dimensional space, making them suitable for applications requiring mobility and adaptability. This mobility enables FANETs to be deployed quickly and flexibly in various scenarios, including disaster areas, search-and-rescue missions, or temporary surveillance operations. The UAVs offer real-time data collecting and transmission capabilities by being able to visit remote or inaccessible sites.

Flying *ad hoc* networks enable UAVs to act as data collectors and relays. The UAVs can integrate various sensors, cameras, and communication devices to collect data from the surrounding environment. This data can include images, videos, environmental parameters, or other sensor readings. The UAVs can then transmit the collected data to other UAVs, ground stations, or cloud servers for further processing or storage. Relaying allows for extending the communication range and ensuring connectivity in scenarios with limited direct communication links. Flying *ad hoc* networks promote collaborative decision-making among UAVs. The UAVs can exchange information, share their situational awareness, and collaborate on tasks such as network routing, task assignment, or mission planning. This collaborative decision-making enhances the efficiency and effectiveness of FANETs by enabling cooperative behaviors, coordinated actions, and collective optimization of network performance.

Some of the major capabilities of FANETs are mentioned below:

- Aerial surveillance and monitoring: FANETs are well-suited for aerial surveillance and monitoring applications. The UAVs with cameras or other sensors may feed live video, detect objects, track, and monitor activities from an aerial viewpoint. Flying *ad hoc* networks enable efficient data collection, analysis, and transmission, supporting tasks such as border surveillance, wildlife monitoring, infrastructure inspection, and crowd monitoring.
- Search-and-rescue operations: FANETs can be deployed in search-and-rescue operations to enhance situational awareness, communication, and coordination. The UAVs equipped with thermal cameras, sensors, or communication systems can assist in locating missing persons, providing aerial views of the search area, and facilitating communication between rescue teams. The dynamic and self-organizing nature of FANETs enables quick deployment and adaptability to change search conditions.
- Disaster response and management: FANETs play a vital role in disaster response and management scenarios. The UAVs can be deployed to assess the extent of damage, monitor the spread of fires or pollutants, and assist in search and rescue operations. Flying *ad hoc* networks enable rapid data collection, real-time communication, and collaborative decision-making, facilitating timely response coordination and resource allocation during critical situations.

- Communication in remote areas: FANETs can provide communication infrastructure in remote or isolated areas where traditional communication infrastructure is absent or damaged. The UAVs equipped with communication devices can form a network and relay communication signals between remote locations and centralized infrastructure. This capacity is very helpful in situations like remote scientific expeditions, disaster-affected areas, and rural areas.
- Swarm intelligence: FANETs can leverage swarm intelligence, where multiple UAVs work together to accomplish complex tasks. Swarm intelligence enables coordinated and distributed behavior among UAVs, leading to improved efficiency, fault tolerance, and adaptability. Flying *ad hoc* networks can utilize swarm intelligence for tasks such as distributed sensing, area coverage, or collaborative exploration.

Flying *ad hoc* networks are self-organizing networks of UAVs that operate in a decentralized manner. They possess features such as autonomy, dynamic topology, mobility, and flexibility, enabling various applications including surveillance, search-and-rescue operations, disaster response, remote communication, and swarm intelligence. Flying *ad hoc* networks harness the capabilities of UAVs to collect and relay data, collaborate on decision-making, and operate in dynamic and challenging environments. Fog computing and FANETs differ from traditional networking approaches in several key aspects. A comparison of fog computing, FANETs, and traditional networking is presented in Table 10.1.

Thus, fog computing and FANETs introduce new paradigms compared to traditional networking approaches. They bring computation closer to the data source, support decentralized and dynamic networks, reduce latency, enhance scalability, and provide mobility and flexibility. These characteristics make fog computing and FANETs well-suited for real-time applications, edge computing, and scenarios where traditional networking approaches face limitations in terms of latency, scalability, or mobility.

10.3 Synergy Between Fog Computing and FANETs

Fog computing and FANETs are highly complementary technologies that, when combined, offer a powerful and synergistic approach to next-generation networking. Here is an analysis of how fog computing and FANETs complement each other. By bringing networking, storage, and

Table 10.1 A comparison of fog, FANETs, and traditional networks.

Metrics	Traditional networking	Fog computing	FANETs
Infrastructure location	Largely uses cloud servers or remote data centers for data processing and storage and is based on a centralized infrastructure.	Brings the computer infrastructure closer to the data sources by extending it to the network's edge. To enable localized data processing and lower latency, it distributes computational resources across edge devices, including routers, gateways, and IoT devices.	This operates in a decentralized manner without relying on fixed infrastructure. The UAVs act as both data sources and relays, enabling communication and data processing in dynamic and distributed environments.
Network topology	Typically have a hierarchical or star topology, where devices connect to a central network node (e.g., switches, routers, or access points) for data routing and communication.	Supports a distributed network topology with interconnected edge devices. It lessens the requirement for data to be sent to a centralized location for processing by enabling direct connectivity between devices at the edge.	Exhibits a dynamic network topology due to the movement of UAVs. The UAVs self-organize and create an *ad hoc* network, enabling direct communication between drones without relying on fixed infrastructure.
Latency and response time	May introduce latency and increased response times due to data transmission to remote data centers or cloud servers for processing and analysis.	Processes data closer to the data source, which lowers latency. Real-time or almost real-time analysis is made possible by this localized processing, which also supports low-latency applications and speeds up reaction times.	Further reduces latency by enabling localized communication and data processing among UAVs. It speeds up reaction times for important processes by doing away with the requirement to transfer data to centralized servers.

(*Continued*)

Table 10.1 A comparison of fog, FANETs, and traditional networks. (*Continued*)

Metrics	Traditional networking	Fog computing	FANETs
Scalability	Can face scalability challenges when accommodating many devices or handling increased data volumes. Scaling requires adding more centralized infrastructure.	Supports scalability by distributing computational resources across edge devices. It enables efficient resource allocation and load balancing, ensuring optimal utilization as the network grows and complexity.	Exhibits inherent scalability due to the decentralized and self-organizing nature of UAVs. As more UAVs join the network, they contribute to the network's capabilities without relying on additional centralized infrastructure.
Mobility and flexibility	Designed primarily for stationary devices and may not efficiently support mobility or adaptability to dynamic environments.	Leverages the mobility and flexibility of edge devices, enabling seamless integration with mobile and IoT devices. Applications that need to process data in real time and make decisions in dynamic, mobile contexts are supported by it.	Takes mobility to the next level as UAVs can move freely in three-dimensional space. A FANET enables agile and adaptable networking capabilities, particularly useful in scenarios requiring aerial surveillance, disaster response, or remote communication.

computing closer to the network's edge, fog computing lowers latency and allows for localized data processing. A FANET, on the other hand, operates in dynamic and distributed environments, leveraging the mobility and flexibility of UAVs. By integrating fog computing with FANETs, the computing capabilities are extended to the aerial domain, allowing UAVs to act as edge devices for data processing and storage. This combination enables efficient and real-time data processing in diverse scenarios, such as disaster response, surveillance, and remote communication.

By processing data closer to the source, fog computing lowers latency and facilitates real-time or almost real-time decision-making. A FANET enhances this capability by providing agile and mobile networking capabilities. The UAVs in FANETs can collect data from various sources, process it locally using fog computing resources, and make time-sensitive decisions in real time. This collaboration between fog computing and FANETs enables immediate actions based on localized data analysis, reducing reliance on centralized processing and enabling faster response times.

Fog computing supports scalability by distributing computational resources across edge devices, ensuring optimal utilization and load balancing. A FANET, with its self-organizing and decentralized nature, provides inherent scalability as more UAVs join the network. By integrating fog computing with FANETs, the combined system can efficiently allocate and manage resources across both terrestrial edge devices and aerial UAVs. This distributed resource management enables dynamic task offloading, efficient data processing, and flexible resource allocation, contributing to improved scalability and adaptability.

A FANET offers inherent redundancy and fault tolerance due to its distributed nature. If a UAV encounters failure or disruption, neighboring UAVs can take over its responsibilities, ensuring continuity of network operations. Fog computing further enhances this redundancy by providing distributed data storage and processing capabilities. Data can be replicated and stored across multiple edge devices, including both terrestrial and aerial nodes. In case of device failure or communication disruption, data can be seamlessly redirected to alternative nodes, ensuring data availability and fault tolerance.

A FANET enables communication and data relaying in remote or inaccessible areas where traditional networking infrastructure is lacking. By integrating fog computing with FANETs, UAVs can serve as mobile edge nodes, extending the coverage and connectivity of fog computing to these areas. This integration allows for localized data processing, analysis, and decision-making in previously unreachable regions. The combination of FANETs and fog computing improves network connectivity, enhances communication capabilities, and enables the deployment of next-generation services in challenging environments.

Thus, fog computing and FANETs are highly complementary technologies that, when combined, create a powerful networking paradigm. Localized and real-time data processing, low-latency decision-making, scalability, redundancy, and improved coverage are all made possible by the integration of fog computing with FANETs. This combination offers significant benefits in various applications, including disaster response, surveillance, remote communication, and real-time analytics, by leveraging the mobility, agility, and distributed nature of FANETs along with the computational capabilities of fog computing at the network edge.

The integration of fog computing and FANETs brings numerous potential benefits and opens up new possibilities for next-generation networking. Here are some of the key benefits of integrating fog computing and FANETs:

- Reduced latency: By bringing processing power closer to the network's edge, fog computing lowers latency and allows for localized data processing. When combined with a FANET, which operates in dynamic and distributed environments, the integration enables real-time or near real-time data analysis and decision-making. This reduced latency is particularly crucial for applications that require quick response times, such as real-time surveillance, emergency response, and autonomous UAV operations.
- Enhanced reliability and fault tolerance: A FANET exhibits inherent fault tolerance due to its decentralized and self-organizing nature. By integrating fog computing, which provides distributed data storage and processing capabilities, the combined system can further enhance reliability and fault tolerance. Data can be replicated across multiple edge devices, including both terrestrial and aerial nodes, ensuring data availability even in the presence of device failures or communication disruptions. This redundancy improves the overall reliability and resilience of the network.
- Increased scalability: A FANET provides scalability by allowing the network to dynamically adapt and accommodate a varying number of UAVs. By integrating fog computing, the scalability is further enhanced. Fog computing enables the distribution of computational resources across edge devices, ensuring efficient resource allocation and load balancing. As the network expands and more UAVs join, fog computing can dynamically allocate resources to accommodate the increasing demands of data processing and analysis, resulting in improved scalability.
- Localized data processing and analytics: Fog computing enables localized data processing and analytics at the edge of the network. By integrating fog computing with FANETs, UAVs in the network can act as edge nodes, performing data processing and analysis tasks in real time. This localized approach reduces the need for data transmission to centralized servers, minimizes bandwidth usage, and allows

for faster insights and decision-making. It is particularly beneficial in scenarios where real-time data analysis is critical, such as in surveillance, environmental monitoring, and disaster response.

- Dynamic task offloading: The integration of fog computing and FANETs allows for dynamic task offloading between UAVs and edge devices. Fog computing enables UAVs to offload computationally intensive tasks to nearby edge nodes, reducing the computational load on the UAVs and conserving their limited resources, such as battery power and processing capabilities. This dynamic task offloading mechanism optimizes resource utilization, improves overall system performance, and extends the operational capabilities of UAVs in FANET.
- Efficient resource utilization: Fog computing enables the efficient utilization of network resources by distributing computational capabilities to edge devices. By integrating fog computing with FANETs, the computing and storage resources available on UAVs can be leveraged effectively. This allows UAVs to send only pertinent data to centralized systems or other UAVs, allowing them to process and filter data locally. Improved network efficiency results from this strategy's reduction of data transmission over the network, optimization of resource usage, and minimization of bandwidth needs.
- Edge-cloud integration: Fog computing, when integrated with FANETs, facilitates seamless integration between edge computing and cloud computing. The UAVs in FANETs can act as both edge devices and mobile cloud nodes. They can perform local data processing and analysis tasks while connecting to centralized cloud servers for additional computational resources or long-term storage. This integration enables the utilization of both edge and cloud resources, providing a hybrid computing environment that leverages the strengths of both approaches.

10.4 Integration Challenges

For fog computing and FANET integration to be implemented successfully, a number of technological issues must be resolved. These are a few of the main obstacles:

- Resource constraints: Both fog computing and FANETs operate in resource-constrained environments. The UAVs in FANETs have limited computational power, memory, and energy resources, and edge devices in fog computing also have limited resources. Integrating the two requires efficient resource management and optimization techniques to ensure effective utilization of available resources. Balancing computational tasks, data storage, and communication requirements while considering resource limitations is a significant challenge.
- Dynamic network topology: A FANET exhibits a dynamic and constantly changing network topology due to the mobility of UAVs. This dynamic nature poses challenges in maintaining stable connections, routing data efficiently, and managing network nodes. Fog computing needs to adapt to the dynamic topology by incorporating mechanisms for seamless handoff, dynamic reconfiguration, and self-organization. Ensuring continuous and reliable connectivity in the face of UAV movement is a complex task.
- Communication and connectivity: A FANET relies on wireless communication for inter-UAV and UAV-to-ground communication. However, wireless communication in the aerial domain is subject to various challenges, including signal interference, limited bandwidth, and high propagation loss. Integrating fog computing with a FANET requires addressing these communication challenges and developing robust communication protocols that can handle intermittent connectivity, maintain reliable data transmission, and adapt to varying network conditions.
- Data synchronization and consistency: Fog computing and FANETs involve distributed data processing and storage across multiple nodes. Ensuring data synchronization and consistency across these distributed nodes is a challenge. The UAVs in FANETs may experience communication

disruptions or may operate in disconnected environments, leading to data inconsistencies. Maintaining data integrity, consistency, and synchronization across edge devices and UAVs while considering the dynamic nature of the network is a technical challenge.

- Security and privacy: Integrating fog computing and FANETs introduces additional security and privacy concerns. The UAVs in FANETs may be susceptible to attacks, unauthorized access, or data breaches. Fog computing nodes at the edge also need to ensure secure data transmission, authentication, and protection against malicious activities. Addressing security and privacy challenges requires robust encryption mechanisms, access control, intrusion detection, and secure communication protocols.
- Task offloading and load balancing: Fog computing offers dynamic task offloading, which transfers computationally demanding jobs from UAVs to edge devices. However, determining the optimal task offloading decisions and load-balancing strategies in a dynamic FANET environment is a complex problem. To guarantee optimal resource usage and overall system performance, effective load balancing techniques must take into account the dynamic nature of UAV movement, available resources, network circumstances, and job needs.
- Quality of service (QoS) guarantees: Specific quality of service standards, such as minimal latency, dependability, and availability, must be met for fog computing and FANETs. Ensuring consistent and predictable QoS guarantees in a dynamic and distributed environment is challenging. Quality of service provisioning mechanisms must be designed to adapt to changing network conditions, prioritize critical tasks, handle resource constraints, and maintain desired service levels across the integrated fog computing and FANET systems.
- Addressing these technical challenges requires innovative solutions and interdisciplinary research in the areas of resource management, network protocols, data synchronization, security, task offloading, and QoS provisioning. Overcoming these challenges will pave the way for the successful integration of fog computing and FANETs, enabling

the realization of their combined potential in next-generation networking applications.

Communication protocols and resource management play crucial roles in the integration of FANETs and fog computing. Let us examine the issues related to these aspects for both FANETs and fog computing:

- Communication protocols for FANETs
    - Dynamic network topology: A FANET's dynamic and self-organizing nature presents challenges in establishing and maintaining communication links between UAVs. Traditional communication protocols designed for static networks may not be suitable. New protocols are needed to handle frequent topological changes, intermittent connectivity, and mobility-induced disruptions. It is necessary to modify or create routing protocols especially for FANETs, such as proactive (like optimized link state routing protocol) or reactive (like *ad hoc* on-demand distance vector routing).
    - Spectrum management: A FANET operates in a shared wireless spectrum, where multiple UAVs may compete for limited bandwidth. Efficient spectrum allocation and management techniques are necessary to avoid interference and ensure fair access to the available spectrum. Cognitive radio techniques, dynamic spectrum access, and interference avoidance mechanisms can help optimize spectrum usage and mitigate congestion-related issues.
    - Scalability and scalable protocols: FANET systems can involve a large number of UAVs operating simultaneously. To manage the growing number of nodes and the corresponding communication overhead, communication protocols need to be scalable. Protocols need to address scalability challenges related to network control, addressing, and efficient dissemination of control and data messages.
- Resource management for fog computing
    - Resource discovery and allocation: Fog computing involves distributed resources across edge devices. To find the computing, storage, and networking resources that are accessible in the fog environment, effective

resource discovery procedures are needed. Algorithms for allocating resources must be created to match jobs with suitable resources according to attributes like availability, capability, and proximity. Dynamic resource allocation techniques that consider the variability of resource availability and workload are crucial.

- Load balancing and task offloading: In fog computing, load balancing aims to distribute tasks efficiently across edge devices to ensure optimal resource utilization and performance. Network circumstances, device capabilities, and the workload at hand are all elements that load-balancing algorithms must take into account. Task offloading decisions play a crucial role in load balancing, determining which tasks are offloaded from UAVs to edge devices. Dynamic task offloading strategies should consider factors like task requirements, network conditions, and device capabilities to achieve load balancing and maximize resource utilization.
- Energy management: Energy efficiency is critical in resource-constrained environments like FANETs and fog computing. To minimize energy usage and extend the battery life of UAVs and edge devices, energy management procedures must be included. To restore energy in mobile devices, this comprises power-aware resource allocation, energy-aware task scheduling, and energy harvesting techniques. Energy-efficient communication protocols, such as duty cycling and sleep-wake scheduling, can also be applied to conserve energy.
- Quality of service management: Applications using fog computing frequently need to meet strict QoS standards for things like throughput, latency, and dependability. Mechanisms for managing resources must take QoS guarantees into account and distribute resources appropriately. For the fog computing environment, application-specific QoS needs may be satisfied by using bandwidth reservation strategies, traffic prioritization methods, and QoS-aware scheduling algorithms.

The creation of novel resource management plans and communication protocols that are adapted to the particular features and needs of fog computing and FANETs is necessary to address these problems. It necessitates interdisciplinary research and collaboration among networking, distributed systems, and optimization communities to ensure efficient and reliable communication, optimal resource utilization, and satisfactory QoS in integrated FANETs and fog computing systems.

10.5 Architectural Implications

The integration of fog computing and FANETs has a significant impact on network architecture, leading to the emergence of a new paradigm that enhances the capabilities of traditional network architecture. Let us examine the key impacts:

- Decentralized architecture: Fog computing integrated with a FANET introduces a decentralized architecture where computational and storage resources are distributed across edge devices, including UAVs. This decentralization brings computing capabilities closer to the network edge, enabling real-time data processing, reduced latency, and improved response times. It shifts the focus from relying solely on centralized cloud infrastructure to leveraging distributed computing resources at the network edge.
- Edge-computing capabilities: Fog computing, when integrated with a FANET, extends the capabilities of edge devices, including UAVs, beyond their traditional networking functions. These edge devices become capable of processing and analyzing data locally, reducing the need for transmitting all data to a centralized cloud. This enables faster decision-making, reduces bandwidth consumption, and improves application performance.
- Dynamic and mobile environment: A FANET's mobility and dynamic network topology introduce new challenges in network architecture. Network architecture needs to adapt to the rapid changes in the network, establish and maintain connections with moving UAVs, and efficiently route data within the network. This dynamic environment requires flexible and adaptive network architecture that can handle mobility, intermittent connectivity, and changing network conditions.
- Hybrid architecture: The integration of fog computing and FANETs leads to a hybrid architecture that combines

centralized cloud infrastructure, fog nodes at the network edge, and UAVs forming a dynamic mesh network. This hybrid architecture leverages the strengths of both fog computing and FANETs, allowing for efficient data processing and storage at the edge, while still utilizing cloud resources for more extensive data analysis, long-term storage, or resource-intensive computations.

- Distributed intelligence: The integration of fog computing with FANETs enables distributed intelligence across the network. Edge devices, including UAVs, can perform localized decision-making, data analytics, and machine learning tasks based on the real-time data they collect. This distributed intelligence reduces the need for frequent communication with centralized systems, enhances autonomy, and enables intelligent and adaptive behavior in the network.
- Redundancy and resilience: Fog computing integrated with FANETs improves the redundancy and resilience of the network architecture. With distributed computing resources, in case of failure or disruption in one part of the network, other edge devices or UAVs can continue to provide services and process data. This redundancy improves network reliability, and fault tolerance, and ensures continuous operations even in challenging environments.
- Scalability and flexibility: The network design gains scalability and flexibility *via* the inclusion of fog computing and FANETs. Horizontal scaling is made possible by the increased computation and storage capacity at the network edge brought about by the additional edge devices and UAVs. The network architecture can adapt to changing requirements, accommodate varying workloads, and seamlessly integrate new devices into the network.

Overall, the integration of fog computing and FANETs introduces a paradigm shift in network architecture, enabling decentralized, edge-centric, and intelligent networks. It brings enhanced performance, improved response times, scalability, and adaptability to meet the evolving needs of next-generation networking applications in various domains such as surveillance, disaster response, and environmental monitoring.

The integration of FANETs and fog computing enhances network resilience and data processing efficiency by leveraging the capabilities of both technologies. This integration achieves the following benefits:

- Network resilience
  - Distributed architecture: By combining a FANET with fog computing, a distributed architecture is introduced, distributing storage and processing resources across UAVs and edge devices. This decentralized method removes single points of failure, which increases network resilience. Even if certain edge devices or UAVs experience disruptions or failures, other devices can continue to provide services and process data, ensuring the network's continuous operation.
  - Redundancy: Redundancy in data processing and storage is made possible by the dispersed nature of fog computing combined with FANETs. Multiple UAVs or edge devices can analyze and store data, lowering the possibility of data loss in the event of a failure. Redundancy enhances network resilience by ensuring that data is not dependent on a single device or location, thereby mitigating the impact of device or link failures.
  - Self-healing capabilities: FANETs and fog computing systems can incorporate self-healing mechanisms to recover from failures or disruptions autonomously. When a failure is detected, the network can dynamically reroute data flows, reassign tasks to alternative devices, and adapt its operations to maintain service continuity. This self-healing capability enhances network resilience and reduces the need for manual intervention.
  - Mobility support: A FANET's mobility, combined with fog computing, enables resilient networking in dynamic environments. As UAVs move and network topology changes, the integration ensures continuous connectivity by dynamically adjusting routing paths and establishing new connections with neighboring devices. This mobility support enhances network resilience in scenarios where traditional networks may struggle to maintain connectivity due to the mobility of devices.
- Data processing efficiency
  - Edge computing: By bringing computational power to the network edge, fog computing makes it possible to process data closer to the source. The UAVs can function as edge computing nodes, processing and analyzing data in real time, by integrating with a FANET.

Through local processing near the point of generation, the integration lowers the delay that results from sending data to a centralized cloud for processing. This facilitates quicker decision-making and increases the effectiveness of data processing.

- Task offloading: Fog computing integrated with a FANET allows for dynamic task offloading from UAVs to nearby edge devices based on computational requirements, network conditions, and energy constraints. This task-offloading mechanism improves data processing efficiency by distributing the workload across available resources, including edge devices and UAVs. It optimizes resource utilization, reduces the burden on UAVs, and improves overall system performance.
- Data filtering and aggregation: A FANET uses several UAVs to produce a massive amount of data. When fog computing and a FANET are combined, data may be filtered and aggregated at the edge, sending only pertinent information to centralized or cloud-based systems for additional examination. By reducing the amount of data that must be transferred across the network, this filtering and aggregation process maximizes bandwidth consumption and boosts the effectiveness of data processing.
- Real-time analytics: At the network edge, real-time data analytics and decision-making are made possible by fog computing combined with a FANET. Real-time analytics may be done on streaming data by utilizing the processing power of edge devices and UAVs. This facilitates timely insights and actionable intelligence, enhancing data processing efficiency for time-critical applications such as surveillance, disaster response, or traffic management.

The integration of a FANET and fog computing enhances network resilience by providing a distributed and redundant architecture, self-healing capabilities, and mobility support. It improves data processing efficiency through edge computing, task offloading, data filtering, and real-time analytics. These benefits contribute to more robust and efficient network operations, enabling reliable and timely data processing in dynamic and resource-constrained environments.

The integration of a FANET and fog computing introduces several considerations for future network design and infrastructure planning.

To guarantee the integrated system's effective deployment and operation, these ramifications must be taken into account. Some of the key considerations are discussed in this section. Future network designs should incorporate a hybrid architecture that combines centralized cloud infrastructure, fog nodes at the network edge, and FANET components such as UAVs. The architecture should be scalable, flexible, and capable of accommodating dynamic network topologies, mobility, and distributed computing capabilities. Deploying fog nodes and other edge computing resources at key locations is a key component of infrastructure planning that will support FANETs and facilitate effective data processing at the network edge. The placement of fog nodes should consider factors such as coverage, connectivity, computational capacity, and power availability to ensure optimal performance and resource utilization. Future network designs should address the challenges of establishing and maintaining connectivity in dynamic and mobile environments. Robust communication protocols, adaptive routing mechanisms, and dynamic spectrum management techniques should be considered to support seamless communication among FANET components and fog nodes.

Infrastructure planning should account for efficient resource management in the integrated system. This entails assigning computing power, storage space, and network bandwidth to meet the computational needs of edge devices, UAVs, and fog nodes. To guarantee efficient resource use and avoid bottlenecks, load balancing, resource allocation algorithms, and resource monitoring methods should be put into place. Security and privacy issues must be given top priority in network architecture and infrastructure development. Secure communication protocols, intrusion detection, authentication, encryption, and other strong security features should all be included in the system. Protecting sensitive data gathered and processed by the integrated system should also take privacy-preserving measures and adherence to privacy laws into account. Future network designs should focus on interoperability and standardization to facilitate the seamless integration of FANET and fog computing components. It is important to provide standard protocols, interfaces, and data formats to guarantee device and system compatibility and interoperability among various suppliers. In multi-vendor setups, this facilitates scalability, seamless integration, and cooperative operations.

Infrastructure planning should address the energy constraints of FANET and fog computing components, especially UAVs with limited battery capacity. Optimizing energy consumption and extending the operational lifespan of UAVs and edge devices may be achieved by including energy-efficient protocols, power management strategies, and renewable energy sources. As the number of edge devices, UAVs, and fog nodes increases, future network solutions must be scalable and maintainable. Network management frameworks, monitoring tools, and automation techniques should be employed to simplify the management and provisioning of resources, configuration updates, and fault detection in the integrated system. Network design and infrastructure planning should consider data governance and compliance requirements. Establishing policies and procedures pertaining to data ownership, access control, data sharing, and legal compliance (such as GDPR) is imperative to guarantee the conscientious management of data gathered and processed by the integrated system. To promote research, innovation, and the creation of standards and best practices for next network architecture and infrastructure planning, cooperation between academic institutions, business, and regulatory agencies is crucial. Collaboration fosters knowledge exchange, addresses emerging challenges, and facilitates the adoption of integrated FANET and fog computing systems in various domains.

Considering these implications in future network design and infrastructure planning ensures the successful deployment, scalability, and long-term viability of the integrated FANET and fog computing system, enabling the realization of its full potential in diverse application scenarios.

10.6 Deployment Considerations

A number of pragmatic factors must be taken into account while deploying fog computing integrated with a FANET to guarantee an effective setup and functioning of the combined system.

- Infrastructure planning
  - Site selection: Identify suitable locations for deploying fog nodes considering factors such as coverage area, connectivity, power availability, and environmental conditions. At the network edge, effective data processing and transmission are ensured by the optimal positioning of fog nodes.
  - Power and connectivity: Ensure reliable power sources and network connectivity for fog nodes, considering factors such as availability, redundancy, and backup solutions. Robust power and connectivity infrastructure is crucial for the uninterrupted operation of fog nodes and their communication with FANET components.

- Device selection and deployment
  - Edge devices: Select appropriate edge devices that meet the computational, storage, and communication requirements of fog computing. Take into account elements like memory, processing speed, storage space, connection interfaces, and fog computing framework compatibility.
  - Unmanned aerial vehicle selection: Choose suitable UAVs for FANETs that can operate in the desired environment, have sufficient payload capacity for fog computing equipment, and meet regulatory requirements. Consider factors such as flight range, endurance, payload capacity, stability, and integration capabilities.
  - Deployment strategy: Plan the deployment strategy for fog nodes and UAVs to ensure adequate coverage, optimal placement, and efficient utilization of resources. Consider factors such as geographical distribution, density of deployments, proximity to data sources, and load balancing.
- Communication and networking
  - Protocols and standards: To ensure smooth connection between fog nodes, edge devices, and FANET components, use the right communication protocols and standards. Consider protocols for device discovery, routing, synchronization, and data exchange that can handle mobility, intermittent connectivity, and dynamic network topologies.
  - Bandwidth and spectrum management: Address bandwidth requirements and spectrum management challenges to ensure efficient data transmission and communication in the integrated system. Consider mechanisms for dynamic bandwidth allocation, spectrum sharing, interference mitigation, and QoS provisioning.
- Security and privacy
  - Security mechanisms: Implement robust security mechanisms to protect the integrated system from threats and attacks. To protect data, devices, and network infrastructure, use strategies including encryption, authentication, access control, intrusion detection, and secure communication protocols.
  - Privacy considerations: Address privacy issues pertaining to the integrated system's data processing, storage,

and gathering. Ensure compliance with privacy regulations, adopt privacy-preserving techniques, and establish policies for data anonymization, user consent, and secure data handling.

- Data management
  - Data processing and analytics: Create effective pipelines for analytics and data processing that make use of fog computing capabilities. Develop methods for real-time analytics, data filtering, and data aggregation. Decide which data should be handled locally at the edge and which data should be sent to the cloud.
  - Data storage and retention: Establish appropriate data storage and retention strategies based on the requirements of the integrated system. Consider factors such as storage capacity, data lifecycle management, data replication, backup strategies, and data synchronization between fog nodes and centralized/cloud storage.
- Monitoring and management
  - Network monitoring: Implement monitoring mechanisms to track the performance, availability, health of fog nodes, edge devices, and FANET components. Monitor network metrics, resource utilization, and communication quality, and detect anomalies or failures for efficient troubleshooting and management.
  - Configuration and updates: Develop mechanisms for easy configuration and updates of fog nodes, edge devices, and FANET components. Consider centralized management frameworks, over-the-air updates, and remote configuration techniques to simplify the management and maintenance of the integrated system.
- Scalability and interoperability
  - Scalability planning: Account for the scalability requirements of the integrated system and design the architecture to accommodate increasing numbers of fog nodes, edge devices, and FANET components. Consider load balancing, resource allocation, and horizontal scaling strategies to ensure scalability.
  - Interoperability: Ensure interoperability between different vendors' devices, protocols, and systems. To facilitate easy integration, teamwork, and scalability throughout

the integrated system, use standard interfaces, protocols, and data formats.

- Regulatory and compliance
  - Regulatory requirements: Consider regulatory requirements and compliance standards specific to the deployment environment. Adhere to aviation regulations, data privacy regulations, spectrum allocation policies, and any other relevant regulations that impact the deployment and operation of the integrated system.
- Testing and validation
  - Testing strategies: Develop comprehensive testing strategies to verify the functionality, performance, and reliability of the integrated system. Test scenarios should include various operating conditions, mobility scenarios, network disruptions, and failure scenarios to ensure the system's robustness.
  - Validation and performance evaluation: Under real-world situations, thoroughly validate the deployed system and assess its performance. Collect performance metrics, analyze system behavior, and assess the system's performance against defined objectives and requirements.

Considering these practical considerations helps ensure the successful deployment, operation, and management of the integrated fog computing and FANET system, enabling efficient data processing, communication, and networking in dynamic and resource-constrained environments. Energy constraints and infrastructure considerations are critical factors for the successful deployment of FANET and fog computing systems. As for energy constraints, UAVs have power limitations. The FANET relies on UAVs, which have limited power capacity due to their small size and weight constraints. Energy-efficient operation is crucial to maximizing the flight endurance and operational duration of UAVs. This includes optimizing power consumption during communication, computation, and sensing tasks.

Investigating energy harvesting strategies can assist in reducing the energy limitations in FANET installations. The power needs of UAVs and fog computing devices can be supplemented by harvesting energy from renewable sources like solar, wind, or kinetic energy, increasing their endurance and lowering their need on batteries. Optimizing energy usage in fog computing devices and UAVs may be achieved by using effective power management methods, including sleep modes, dynamic voltage scaling, and job scheduling algorithms. Balancing the trade-off between performance and power consumption is crucial to prolonging the operational time and ensuring the reliability of the integrated system. Regarding infrastructure considerations, it is important to strategically place fog nodes inside the FANET coverage region to provide effective data processing and communication. Installing fog nodes with the processing power, storage, and connection to enable real-time data analysis and network edge decision-making should be part of infrastructure design. The smooth functioning of FANET and fog computing systems depends on dependable connectivity.

To guarantee reliable and continuous connectivity between fog nodes, UAVs, and other network components, the deployment of communication infrastructure, such as base stations, antennae, and network gateways, should be taken into consideration. Fog nodes in the network require backhaul connectivity to transmit and receive data to and from the cloud or centralized servers. The availability of high-speed, low latency back-haul connections, such as fiber optic or wireless links, is crucial to support the data transmission requirements and ensure effective integration of fog computing and FANETs. Designing a scalable and redundant infrastructure is important to accommodate the growing number of edge devices, fog nodes, and UAVs in the network. The infrastructure should support horizontal scaling, load balancing, and fault tolerance to ensure uninterrupted operation and handle increasing data processing and communication demands. Fog computing devices may have limited computational resources and storage capacity. Consideration should be given to resource-constrained environments and the efficient utilization of available resources. Techniques such as edge caching, distributed computing, and offloading of computational tasks can help optimize resource usage and improve system performance. Flying *ad hoc* networks and fog computing deployments should consider environmental factors such as temperature, humidity, and weather conditions. The infrastructure should be designed to withstand and operate in harsh environmental conditions to ensure the reliability and longevity of the integrated system.

Addressing energy constraints and infrastructure considerations is crucial for the successful deployment of FANETs and fog computing systems. Optimizing energy usage, leveraging energy harvesting techniques, deploying suitable edge infrastructure, ensuring robust connectivity, and considering scalability and redundancy are key factors in building an efficient and sustainable integrated system. By carefully managing energy resources and infrastructure, the deployment can achieve reliable and long-lasting performance in diverse application scenarios.

This chapter provides a thorough examination of the integration of fog computing with FANETs, emphasizing the advantages that might arise, as well as the practical, technological, and benefit-related issues. The chapter's main conclusions and contributions may be summed up as follows:

- Synergy between fog computing and FANETs: A strong foundation for next-generation networking is created by the combination of fog computing and FANETs. It enables efficient data processing and communication at the network edge, leveraging the mobility and flexibility of UAVs in FANET deployments.
- Benefits of integration: Reduced latency, increased scalability, strengthened network resilience, effective data processing and analytics, and support for real-time applications are just a few advantages of the integration. It enables dynamic resource allocation, localized decision-making, and effective utilization of edge resources.
- Technical challenges: The chapter examines various technical challenges in integrating fog computing and FANETs. These challenges include energy constraints, infrastructure considerations, communication protocols, resource management, security, coordination, and interoperability. Addressing these challenges is crucial for the successful implementation and operation of the integrated system.
- Practical considerations: The chapter provides insights into practical considerations for deploying fog computing integrated with FANETs. It covers aspects such as infrastructure planning, device selection and deployment, communication and networking protocols, security and privacy considerations, data management, monitoring and management, scalability, and interoperability. Considering these factors ensures the effective deployment and operation of the integrated system.
- Regulatory and compliance aspects: The chapter highlights the importance of examining regulatory aspects and compliance requirements related to FANET and fog computing deployments. It emphasizes adherence to aviation regulations, data privacy regulations, spectrum regulations, security standards, and ethical frameworks to ensure the legal and ethical operation of the integrated system.

- Network resilience and data processing efficiency: Through the facilitation of localized data processing and decision-making, the combination of fog computing and FANETs improves network resilience. It enhances overall system performance, lessens the effect of network outages, and decreases dependency on centralized cloud resources. Using fog computing capabilities, the chapter highlights the significance of effective data processing and analytics at the network edge.
- Implications for future network design: Future network architecture and infrastructure planning will be significantly impacted by the combination of fog computing with FANET. It calls for scalable and flexible architecture, efficient resource management mechanisms, and advanced communication protocols to support the evolving needs of next-generation networking.

There is a lot of potential for additional studies and development on the integration of fog computing and FANETs. Future directions for this field's research and difficulties include:

- Resource optimization: Investigating advanced resource optimization techniques for fog-computing integrated FANETs is crucial. To maximize resource use and improve system performance, this entails creating effective models and algorithms for job offloading, resource allocation, load balancing, and energy management.
- Mobility and handover management: Addressing the challenges related to the mobility of UAVs in FANETs is an important research direction. Developing seamless handover mechanisms, efficient mobility management protocols, and trajectory planning algorithms to ensure uninterrupted communication and data processing during UAV movement will be crucial.
- Communication protocols: Designing and optimizing communication protocols specifically tailored for fog computing-integrated FANETs is essential. The development of robust and effective protocols for spectrum sharing, adaptive routing, low-latency and dependable communication, and interference control in dynamic and resource-constrained systems might be the focus of research efforts.

- Security and privacy: Improving the privacy and security of the integrated fog-computing FANET is a major task. To safeguard data and guarantee safe and reliable communication inside the integrated system, future research should concentrate on creating reliable authentication methods, encryption techniques, intrusion detection and prevention systems, and privacy-preserving algorithms.
- Edge intelligence and AI integration: It is very promising to investigate how edge intelligence and AI methods may be used with fog computing-integrated FANETs. Research may look at the use of AI algorithms at the network edge, which can help with autonomous UAV coordination, intelligent decision-making, anomaly detection, and predictive analytics for better situational awareness and system performance.
- Scalability and heterogeneity: Scalability remains a challenge in fog computing-integrated FANETs, particularly when dealing with a large number of UAVs and heterogeneous edge devices. Future research can focus on developing scalable architectures, distributed algorithms, and adaptive mechanisms to handle the increasing scale and heterogeneity of the integrated system.
- Standardization and interoperability: It is essential to establish standards and protocols for compatibility across various FANET systems and fog computing platforms. Future research should aim to develop standard interfaces, data formats, and communication protocols to ensure seamless integration, compatibility, and collaboration between different vendors and technologies.
- Testbeds and real-world deployments: Building testbeds and conducting real-world deployments of fog-computing integrated FANET systems is an important area of research. It allows for practical validation, performance evaluation, and the identification of challenges and optimizations in real-world scenarios.
- Application-specific studies: Application-specific research will shed light on the benefits and practical application of the integrated system as it explores the effects of fog computing-integrated FANETs in a variety of domains, including industrial automation, smart cities, precision agriculture, and disaster response.

10.7 Conclusion

This chapter advances our understanding of fog computing and FANET integration in a thorough way. It lists the possible advantages, looks at the technological difficulties, offers useful advice, and emphasizes the consequences for network design going forward. The chapter's conclusions and ideas are a great resource for academics, professionals, and decision-makers working in the subject of next-generation networking. The topic of fog computing-integrated FANETs will evolve further by addressing the problems and future research directions. This will make it possible to produce scalable, dependable, and efficient next-generation networking solutions for a variety of applications.

References

1. Tadakamalla, U. and Menascé, D.A., Autonomic Resource Management for Fog Computing. *IEEE Trans. Cloud Comput.*, 10, 4, 2334–2350, 1 Oct.-Dec. 2022, doi 10.1109/TCC.2021.3064629.
2. Kaur, A., Auluck, N., Rana, O., Real-Time Scheduling on Hierarchical Heterogeneous Fog Networks. *IEEE Trans. Serv. Comput.*, 16, 2, 1358–1372, 1 March-April 2023, doi: 10.1109/TSC.2022.3155783.
3. Yu-Jie, S., Hui, W., Cheng-Xiang, Z., Balanced Computing Offloading for Selfish IoT Devices in Fog Computing. *IEEE Access*, 10, 30890–30898, 2022, doi: 10.1109/ACCESS.2022.3160198.
4. Sun, L., Xue, G., Yu, R., TAFS: A Truthful Auction for IoT Application Offloading in Fog Computing Networks. *IEEE Internet Things J.*, 10, 4, 3252–3263, 15 Feb.15, 2023, doi: 10.1109/JIOT.2022.3143101.
5. Aazam, M., Islam, S.U., Lone, S.T., Abbas, A., Cloud of Things (CoT): Cloud-Fog-IoT Task Offloading for Sustainable Internet of Things. *IEEE Trans. Sustain. Comput.*, 7, 1, 87–98, 1 Jan.-March 2022, doi: 10.1109/TSUSC.2020.3028615.
6. Wu, Q., *et al.*, Routing protocol for heterogeneous FANETs with mobility prediction. *China Commun.*, 19, 1, 186–201, Jan. 2022, doi: 10.23919/JCC.2022.01.014.
7. Beegum, T.R., Idris, M.Y.I., Ayub, M.N.B., Shehadeh, H.A., Optimized Routing of UAVs Using Bio-Inspired Algorithm in FANET: A Systematic Review. *IEEE Access*, 11, 15588–15622, 2023, doi: 10.1109/ACCESS.2023.3244067.
8. Zhu, L., Karim, M.M., Sharif, K., Xu, C., Li, F., Traffic Flow Optimization for UAVs in Multi-Layer Information-Centric Software-Defined FANET. *IEEE Trans. Veh. Technol.*, 72, 2, 2453–2467, Feb. 2023, doi: 10.1109/TVT.2022.3213040.

9. Li, J., Ye, D.H., Kolsch, M., Wachs, J.P., Bouman, C.A., Fast and Robust UAV to UAV Detection and Tracking from Video. *IEEE Trans. Emerg. Top. Comput.*, 10, 3, 1519–1531, 1 July-Sept. 2022, doi: 10.1109/TETC.2021.3104555.
10. Valianti, P., Kolios, P., Ellinas, G., Energy-Aware Tracking and Jamming Rogue UAVs Using a Swarm of Pursuer UAV Agents. *IEEE Syst. J.*, 17, 1, 1524–1535, March 2023, doi: 10.1109/JSYST.2022.3179632.
11. B. S., A., Jadhav, H., Kumaravelu, V.B., Soria, F.R.C., Sayeed, M.S., Murugadass, A., Smart Reconfigurable Intelligent Surface with Discrete Phase Shifter for Next Generation Networks. *2022 International Conference on Wireless Communications Signal Processing and Networking (WiSPNET)*, Chennai, India, pp. 178–182, 2022, doi: 10.1109/WiSPNET54241.2022.9767157.
12. Gross, P. *et al.*, A National Programmable Infrastructure to Experiment with Next-Generation Networks. *2021 IFIP/IEEE International Symposium on Integrated Network Management (IM)*, Bordeaux, France, pp. 778–782, 2021.

11

Quantum Cryptography and FANET Security

M. G. Sumithra[1], R. Remya[2*], Ashwini A.[3], G. Dhivyasri[4] and M. Manikandan[5]

[1]*Dept. of ECE, Sri Krishna College of Technology, Coimbatore, India*
[2]*Dept. of ECE, Sri Krishna College of Engineering and Technology, Coimbatore, India*
[3]*Vel Tech Rangrajan Dr. Sagunthala R&D Institute of Science and Technology, Chennai, India*
[4]*Dept. of CSE (Data Science), Sai Vidya Institute of Technology, Bengaluru, India*
[5]*Dept. of ECE, Presidency University, Bengaluru, India*

Abstract
Many cryptographic techniques were available to achieve a secure communication. Because of the open nature of wireless transmission, the communication is not regarded secure. Moreover, with the arrival of quantum computing technology, it overcomes the issues faced during the communication. Quantum cryptography uses the principle of quantum physics for secured information processing. Such processed information is transmitted from one end to the other end. It is regarded as a necessary one, if the communication channel cannot be trusted. Furthermore, a flying *ad hoc* network (FANET) is an *ad hoc* network, which rectifies the problem associated with communication among the ground station and unmanned ariel vehicles. It is supportive of real-time data transfer; also, it holds security features of an *ad hoc* network.

Keywords: FANET, *ad hoc* network, quantum physics, cryptography

11.1 Introduction

Quantum cryptography has the property of quantum mechanics to transmit the data securely without hacking. In general, cryptography is the method to encrypt and protect the data with some secret key. So, the person at the other end who has the same secret key can decrypt the data.

**Corresponding author*: remiamernath@gmail.com

Taskeen Zaidi, Adarsh Kumar, Saurav Mallik and Keshav Kaushik (eds.) Autonomous Flying Ad-Hoc Networks, (233–244) © 2025 Scrivener Publishing LLC

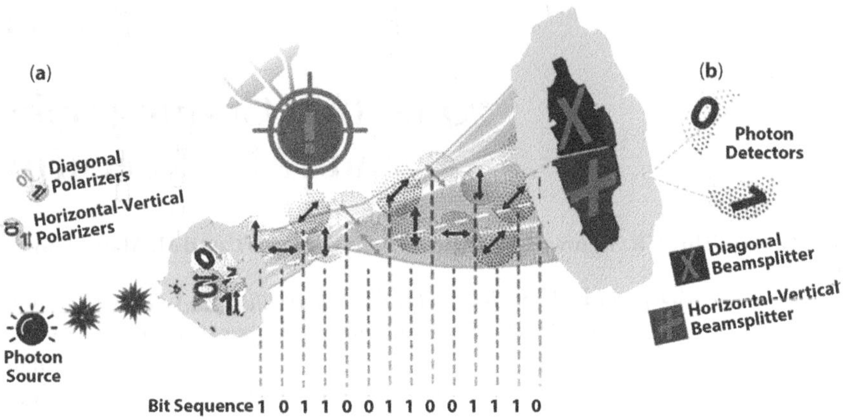

Figure 11.1 Quantum transmission. Source: https://quantumxc.com/blog/quantum-cryptography-explained/.

Quantum cryptography is entirely different from the traditional cryptographic approach. It depends on physics instead of mathematics. Without prior knowledge of the sender's secret key, it can encrypt the data. For encryption, it uses a quantum state to input data. The content will not be sent to either the sender nor the receiver. Also, it is impossible to view the data from one end to the other end. The transmission media used here is the optical fiber. The representation of photon is in binary form [1]. The secrecy of the information is based on quantum mechanics, and it is depicted in Figure 11.1.

Its security depends on the following three properties:

- Particles may exist in more than one place or state at a time.
- Properties of quantum cannot be observed without disturbing them.
- Its whole property could not be copied.

11.1.1 How Does It Work?

In quantum cryptography, photons were used. They acted as information carriers in optical fiber cables. For key exchange, quantum key distribution (QKD) is regarded as the current quantum cryptography. It was developed in the year 1984 with a model. In that model, it assumes A and B wish to share the information more securely. 'A' initiates the transmission by sending a key to 'B.' The key is nothing but a sequence of photons, which travels in the same direction. Each photon corresponds to one bit of data; it may

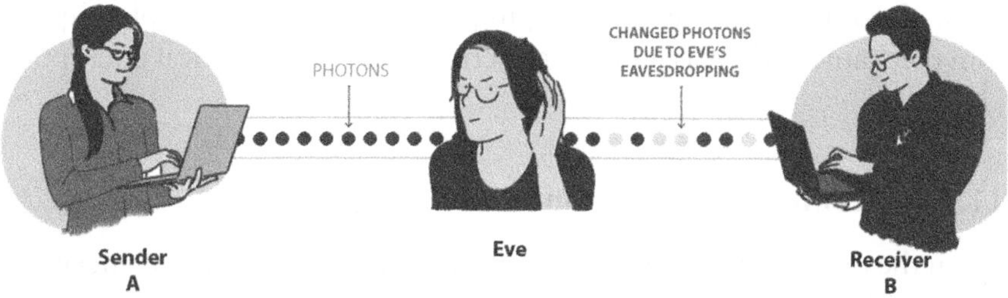

Figure 11.2 Quantum cryptography modeling. Source: https://www.techtarget.com/searchsecurity/definition/quantum-cryptography.

be either 0 or 1. In addition to this, the photons were oscillating in one direction. Before the data transmission, the photon was allowed to pass through a polarizer. A polarizer acts like a filter that allows only photons with the same oscillation effect and makes the remaining photons pass in a changed state of vibration. The polarized state may be in horizontal, vertical, 45° left or 45° right angles. Such a filtered photon then travels through the optical fiber to reach 'B.' It uses a beam splitter to read the photons' polarization. At the receiving end, it is not aware of the exact polarization of the received photon, so 'B' choses one polarization randomly. Now 'A' compares 'B' to what she used to polarize the secret key and makes 'B' transmit each photon. 'B' afterwards confirms whether he/she used the exact polarizer. The wrong photon coming out of the splitter has to be discarded; while the other photon sequence was considered as the key. If Eve tries to decode, then the polarization must be done by Eve, which creates a discrepancy between 'A' and 'B'. Since Eve has the facility to listen to the content, which was shared by 'A.' But 'B' has the advantage to communicate with 'A.' Communicating regarding the usage of polarizer was used by each photon. Both 'A' and 'B' know about the involvement of 'Eve' between the sender and the receiver during the data transmission. This information is not known by the 'Eve' [2], and it is given in Figure 11.2.

11.1.2 Difference Between Post-Quantum Cryptography and Quantum Cryptography

Post-quantum cryptography utilizes a public key encryption algorithm to secure data against the attack by a quantum computer. If the same method has been used by traditional computers, it means it took a lot of time to break the post-quantum cryptography. Since quantum computers utilizes Shor's algorithm, it breaks the math-based systems in a moment.

On the other hand, quantum cryptography utilizes the principle of quantum mechanics, whereas it does not perform any mathematical encryption [3].

11.1.3 Tomorrow's Solution

Upon the development of quantum computers, the assurance of encrypted data is risky. Quantum cryptography through QKD provides a solution to protect the information [4].

11.1.4 Flying *Ad Hoc* Network Security

The usage of wireless networks is prone to the attack of Eve and other interference attacks. Unmanned aerial vehicles (UAVs), usually referred to as drones, can communicate with one another using a mobile *ad hoc* network (MANET) called a flying *ad hoc* network (FANET). It is also known as an aerial *ad hoc* network or a drone *ad hoc* network. It is a UAV or drone-specific wireless communication network that is dynamic and decentralized. A FANET is a network made up of UAVs that can communicate wirelessly with one another without the use of permanent networks like base stations. These networks are suitable for applications including search-and-rescue operations, surveillance, precision farming, and military operations due to their scalability, mobility, and resource-constrained nature. When it comes to maximizing energy use in UAVs with constrained resources, FANETs pose special difficulties. Flying *ad hoc* networks are a subset of MANETs that have been modified to meet the needs and constraints of coordinated UAV flight [6].

11.2 Components of a FANET

11.2.1 Unmanned Aerial Vehicles

In a FANET, UAVs serve as the main nodes. These can be multirotor drones, fixed-wing aircraft, or any other kind of aerial vehicle with networking and communication capabilities. Unmanned aerial vehicles are aerial platforms that can fly without a human pilot within them. Multi-rotor drones like hexacopters, octocopters, quadcopters, Fixed-wing aircraft, and other specialized designs are of many shapes, sizes, and configurations. The precise mission and operating requirements of the FANET determine the type of UAV that should be used.

11.2.2 Communication Hardware

To link with other UAVs within its range, each UAV in a FANET is outfitted with wireless networking hardware, such as radio transceivers. These communication tools give UAVs the ability to connect to other UAVs in the network and maintain that connection, creating an airborne mesh system. Depending on the application and operational range, communication can take place across radio frequencies using Wi-Fi, Bluetooth, or other long-range communication protocols.

11.2.3 Routing Protocol

Flying *ad hoc* networks use specialized routing protocols designed for the decentralized and dynamic nature of UAV networks. These protocols control data transmission between UAVs and can change to accommodate shifting network topologies.

11.2.4 Mobility

One of the distinguishing features of UAVs is their mobility. They have the ability to quickly change locations and maneuver across airspace. Since UAVs must adapt to shifting network topologies, follow assignments, and react to changing environmental circumstances, mobility is crucial in FANETs [5].

11.2.5 Sensors and Payloads

Unmanned aerial vehicles frequently carry a variety of sensors and payloads tailored to their mission objectives. Cameras for taking pictures and recording videos, environmental sensors for gathering data, thermal cameras for use in search-and-rescue missions, LiDAR for mapping, and even cargo spaces for use in transport missions might all fall under this category. The payload or data gathered contribute to the FANET's overall functionality.

11.2.6 Autonomy and Control

To enable autonomous flying, UAVs are frequently fitted with onboard CPUs, GPS locators, and detectors like accelerometers and gyroscopes. They may occasionally be managed remotely by people working out of ground control stations. Different levels of autonomy are possible, from straightforward waypoint navigation to sophisticated obstacle-avoiding and capacity for decision-making.

11.2.7 Energy Considerations

Unmanned aerial vehicles have a limited amount of onboard energy sources, typically in the form of batteries that can be recharged. Their flight time and range are constrained by this low energy capacity. To maximize the operational time of UAVs, FANETs must employ energy-efficient messaging and navigation techniques.

11.2.8 Ground Control Station

A ground control station acts as a hub for managing and directing the network's UAVs. It can communicate with UAVs that transmit and receive data between them.

11.3 Characteristics of FANETS

11.3.1 Mobility

As UAVs in a FANET are so mobile, the topology of the network is continually changing as UAVs move about. To react to quick topology changes, routing systems must be robust due to their dynamic nature.

11.3.2 Decentralization

Since FANETs often lack permanent components like base stations and access points, they are classified as decentralized networks. The UAVs can interact with one another directly or *via* many hops.

11.3.3 Scalability

Flying *ad hoc* networks are useful for applications including surveillance, search-and-rescue operations, and environmental monitoring since they can scale to incorporate a large number of UAVs.

11.3.4 Constraint to Resources

Unmanned aerial vehicles frequently have constrained computing, storage, and power resources. Flying *ad hoc* network protocols must be capable of operating on hardware with limited resources and be energy-efficient.

11.4 Challenges in FANETS

- Due to the UAVs' mobility, the network topology is continually changing, which might cause frequent link interruptions. Routing protocols must effectively manage this.
- Unmanned aerial vehicles flying close to one another may face interference, and the communication range may be constrained, particularly for UAVs at higher altitudes.
- Spoofing, jamming, and eavesdropping are just a few of the security risks that FANETs are vulnerable to. Network integrity and data confidentiality must be guaranteed.
- As UAV batteries have a finite lifespan, it is crucial to employ energy-efficient transmission and routing techniques to extend their operational window.

11.5 Applications of FANETS

11.5.1 Search and Rescue

To conduct aerial surveillance over a vast search area, FANETs allow the deployment of several UAVs outfitted with cameras with excellent resolutions and sensors. This enables search-and-rescue personnel to quickly locate missing people, or stranded hikers by giving them access to real-time aerial photos and video feeds. In disaster-stricken areas, FANETs can be used to look for survivors, communicate information, and organize rescue operations. To ensure the protection of survivors or injured people during transportation to medical facilities or safe zones, FANET-equipped UAVs can help direct and monitor the evacuation of those people. The applications of quantum cryptography are given in Figure 11.3.

Surveillance and Monitoring:
Large areas, such forests, borders, and wildlife habitats, are monitored using FANETs. They can keep an eye on wildlife and the environment. Flying *ad hoc* networks enable the synchronized coordination of several UAVs. They can work together to conduct simultaneous searches in various places and cover greater areas more effectively.

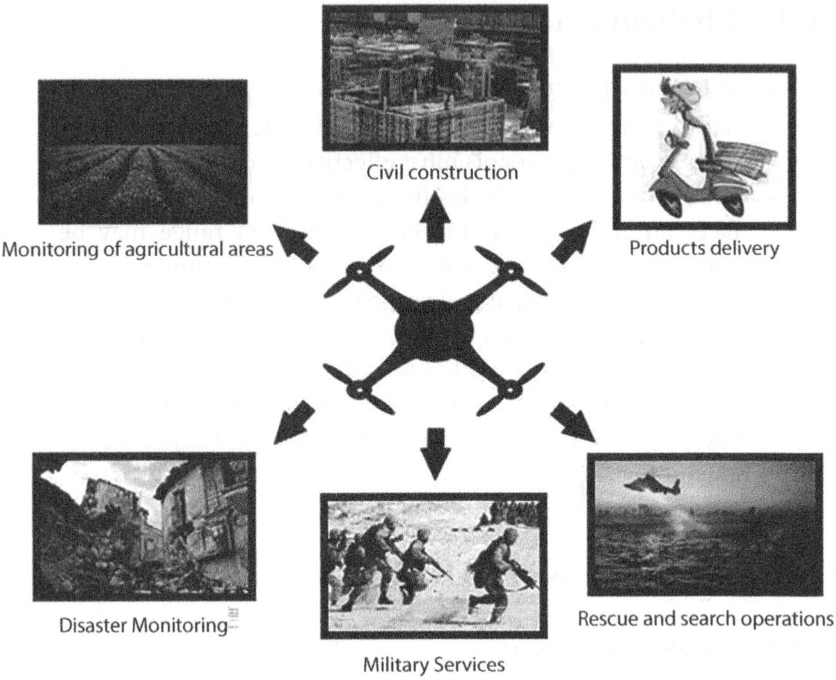

Figure 11.3 Flying ad hoc network applications. Source: https://www.intechopen.com/chapters/67574.

11.5.2 Precision Agriculture

Unmanned aerial vehicles in FANETs can be utilized in agriculture for pesticide application, soil analysis, and crop monitoring. Inaccessible locations, such as fallen structures, disaster-affected areas, or dangerous situations, can be explored and assessed by UAVs. They can offer insightful information about the state of the structures and the presence of survivors.

11.5.3 Communication Relays

When regular networks are inaccessible in rural or disaster-affected locations, FANETs might offer temporary communication infrastructure. Response times to emergencies are shortened with FANETs because UAVs can be quickly deployed from a central command station or even from a mobile base. This flexibility is essential in circumstances where time is of the essence, like searching for people who have gone missing in far-off places.

11.5.4 Military and Defense

Flying *ad hoc* networks are used for communication on the battlefield, intelligence gathering, and reconnaissance. After a tragedy, FANETs can continue to be used to assess damage, identify areas that need to be rescued immediately, and conduct additional searches for any survivors who may still be alive. Real-time data from UAVs can give search-and-rescue personnel current knowledge of shifting weather patterns, topographic challenges, and potential hazards, improving overall situational awareness.

11.6 Quantum Computing in FANET Security

In FANETs, quantum computing plays a major role in keeping the information more secure than other communication and cryptographic applications. Here, communication in every module should be done through quantum computing [7]. The overall processing speed is high. Performance-wise, quantum computing is regarded as a better choice. The overall processing speed took time to gather the service from the hub. Once it received the service, it performed the operation sooner than others in terms of service and speed. Its pictorial representation is given in Figure 11.4.

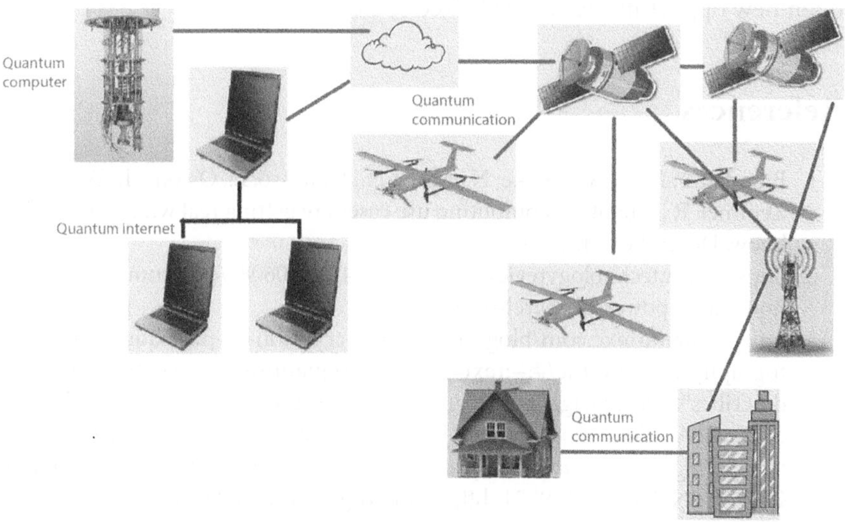

Figure 11.4 Quantum computing in FANET security.

Quantum computing has a footpath in all areas including surveillance, cryptography, image processing, radar, navigation, and chemical detection. Another major drawback of utilizing quantum computing is the high utilization cost. However, security-wise, quantum computing plays a vital role.

Conclusion

Flying *ad hoc* networks with quantum computing are specialized MANETS designed for UAVs. They offer unique capabilities and face challenges related to mobility, decentralization, and resource constraints. A FANET has a wide range of applications, including search and rescue, surveillance, agriculture, and military operations, and their development continues to advance as drone technology evolves. It must manage the dynamic network topology brought on by the mobility of UAVs, optimize energy use in areas with limited resources, and deal with issues about security and privacy. Continuous research, improvements in communication protocols, and the incorporation of cutting-edge technology are necessary for the creation and implementation of FANETs.

Flying *ad hoc* networks with quantum computing technology are positioned to play a more critical role in improving situational awareness, expanding communication coverage, and delivering vital data in various disciplines as UAV technology continues to advance and become more widely available. It has the potential to address complicated problems and open new opportunities in our linked society.

References

1. Biondi, M., Heid, A., Henke, N., Mohr, N., Pautasso, L., Ostojic, I., Wester, L., Zemmel, R., Quantum computing use cases are getting real what you need to know, December 14, 2021.
2. https://www.technologyreview.com/2017/04/05/106085/quantum-computing-now-has-a-powerful-search-tool/.
3. https://quantumxc.com/blog/quantum-encryption-vs-post-quantum-cryptography-infographic/#:~:text=While%20quantum%20cryptography%20describes%20using,against%20an%20attack%20by%20a.
4. National Academies of Sciences, Engineering, and Medicine, *et al.*, *Quantum Computing: Progress and Prospects*, National Academies Press, Washington, 2019, ISBN 9780309479721. https://doi.org/10.17226/25196.

5. Chriki, A., Touati, H., Snoussi, H., Kamoun, F., FANET: Communication, mobility models and security issues. *Comput. Netw.*, 163, 1–10, 2019, https://doi.org/10.1016/j.comnet.2019.106877.
6. Bekmezci, I., Sahingoz, O.K., Temel, S., Flying Ad-Hoc Networks (FANETs): A survey. *Ad Hoc Netw.*, 11, 3, 1254–1270, 2013.
7. Krelina, M., Quantum technology for military applications. *EPJ Quantum Technol.*, 8, 1–53, 2021, 10.1140/epjqt/s40507-021-00113-y.

About the Editors

Taskeen Zaidi, PhD works in the School of Computer Science and Information Technology at Jain University with over 12 years of research and teaching experience. She has published over 40 research papers in international journals, conferences, and workshops. Her areas of interest include cloud computing, ad-hoc networks, distributed computing, and mobile application development.

Adarsh Kumar, PhD is an associate professor in the School of Computer Science at the University of Petroleum and Energy Studies. He has published over 120 research papers in reputed journals, conferences, and workshops. His primary research interests include cybersecurity, cryptography, network security, and ad-hoc networks.

Saurav Mallik, PhD is a research scientist in the Department of Pharmacology and Toxicology at the University of Arizona. He has edited one book and coauthored over 82 research papers in peer-reviewed international journals, conferences, and book chapters. His research areas include data mining, computational biology, bioinformatics, biostatistics, and machine learning.

Keshav Kaushik, PhD is an assistant professor in the Systemic Cluster under the School of Computer Science at the University of Petroleum and Energy Studies with over eight years of teaching experience. He has edited over ten books and published over 65 research papers in international journals and conferences. His research focuses on cybersecurity, digital forensics, and the Internet of Things.

Index

Also of Interest

Check out these other related titles from Scrivener Publishing

AI for Autonomous Vehicles: The Future of Driverless Technology, Edited by Sathiyaraj Rajendran, Munish Sabharwal, Yu-Chen Hu, Rajesh Kumar Dhanaraj, and Balamurugan Balusamy, ISBN: 9781119847465. This title presents the research prospects and security challenges present in the exciting new field of autonomous vehicles by comprehensively introducing advanced concepts of artificial intelligence (AI) in driverless technology.

AUTONOMOUS VEHICLES VOLUME 1: Using Machine Intelligence, Edited by Romil Rawat, A. Mary Sowjanya, Syed Imran Patel, Varshali Jaiswal, Imran Khan, and Allam Balaram. ISBN: 9781119871958. Addressing the current challenges, approaches and applications relating to autonomous vehicles, this groundbreaking new volume presents the research and techniques in this growing area, using Internet of Things, Machine Learning, Deep Learning, and Artificial Intelligence.

AUTONOMOUS VEHICLES VOLUME 2: Smart Vehicles for Communication, Edited by Romil Rawat, Purvee Bhardwaj, Upinder Kaur, Shrikant Telang, Mukesh Chouhan, and K. Sakthidasan Sankaran, ISBN: 9781394152254. The companion to *Autonomous Vehicles Volume 1: Using Machine Intelligence*, this second volume in the two-volume set covers intelligent techniques utilized for designing, controlling and managing vehicular systems based on advanced algorithms of computing like machine learning, artificial Intelligence, data analytics, and Internet of Things with prediction approaches to avoid accidental damages, security threats, and theft.

CONVERSATIONAL ARTIFICIAL INTELLIGENCE, Edited by Romil Rawat, Rajesh Kumar Chakrawarti, Sanjaya Kumar Sarangi, Piyush Vyas, Mary Sowjanya Alamanda, Kotagiri Srividya, and K. Sakthidasan Sankaran. ISBN: 9781394200566. This book presents the need for, design, and application of conversational artificial intelligence (AI). Expert knowledge is shared on leading innovations in natural language processing (NLP) and machine learning (ML) techniques that are frequently combined with more traditional, static kinds of interactive technology, such as chatbots, to create conversational AI.

QUANTUM COMPUTING IN CYBERSECURITY, Edited by Romil Rawat, Rajesh Kumar Chakrawarti, Sanjaya Kumar Sarangi, Jaideep Patel, and Vivek Bhardwaj, ISBN: 9781394166336. This cutting-edge new volume provides a comprehensive exploration of emerging technologies and trends in quantum computing and how it is used in cybersecurity, covering everything from artificial intelligence to how quantum computing can be used to secure networks and prevent cyber crime.

ROBOTIC PROCESS AUTOMATION, Edited by Romil Rawat, Rajesh Kumar Chakrawarti, Sanjaya Kumar Sarangi, Rahul Choudhary, Anand Singh Gadwal, and Vivek Bhardwaj. ISBN: 9781394166183. Presenting the latest technologies and practices in this ever-changing field, this groundbreaking new volume covers the theoretical challenges and practical solutions for using robotics across a variety of industries, encompassing many disciplines, including mathematics, computer science, electrical engineering, information technology, mechatronics, electronics, bioengineering, and command and software engineering.

CONVERGENCE OF CLOUD WITH AI FOR BIG DATA ANALYTICS: Foundations and Innovation, Edited by Danda B. Rawat, Lalit K Awasthi, Valentina Emilia Balas, Mohit Kumar and Jitendra Kumar Samriya, ISBN: 9781119904885. This book covers the foundations and applications of cloud computing, AI, and Big Data and analyses their convergence for improved development and services.

SWARM INTELLIGENCE: An Approach from Natural to Artificial, By Kuldeep Singh Kaswan, Jagjit Singh Dhatterwal and Avadhesh Kumar, ISBN: 9781119865063. This important authored book presents valuable new insights by exploring the boundaries shared by cognitive science, social psychology, artificial life, artificial intelligence, and evolutionary computation by applying these insights to solving complex engineering problems.

FACTORIES OF THE FUTURE: Technological Advances in the Manufacturing Industry, Edited by Chandan Deep Singh and Harleen Kaur, ISBN: 9781119864943. The book provides insight into various technologies adopted and to be adopted in the future by industries and measures the impact of these technologies on manufacturing performance and their sustainability.

AI AND IOT-BASED INTELLIGENT AUTOMATION IN ROBOTICS, Edited by Ashutosh Kumar Dubey, Abhishek Kumar, S. Rakesh Kumar, N. Gayathri, Prasenjit Das, ISBN: 9781119711209. The 24 chapters in this book provide a deep overview of robotics and the application of AI and IoT in robotics across several industries such as healthcare, defense. education, etc.

SMART GRIDS AND INTERNET OF THINGS, Edited by Sanjeevikumar Padmanaban, Jens Bo Holm-Nielsen, Rajesh Kumar Dhanaraj, Malathy Sathyamoorthy, and Balamurugan Balusamy, ISBN: 9781119812449. Written and edited by a team of international professionals, this groundbreaking new volume covers the latest technologies in automation, tracking, energy distribution and consumption of Internet of Things (IoT) devices with smart grids.

Printed and bound by CPI Group (UK) Ltd, Croydon, CR0 4YY
17/09/2026
14964261-0002